THE COMPLETE

HYDRAULIC RAM

MANUAL
THIRD EDITION

Written by:
Tom Moates

Illustrated by:
Chris Legg

Third Edition
Copyright 2024 Tom Moates

The information in this book is true and complete to the best of our knowledge, and all recommendations are made without guarantee on the part of the author or The Homestead Press. The author and publisher disclaim any liability in connection with the use of this information.

Written by Tom Moates
Illustrated by Chris Legg

Printed and manufactured in the U.S.A.

The Homestead Press

Published by:

THE HOMESTEAD PRESS, LLC

P.O. Box 172
Elbert, CO 80106

THEHOMESTEADPRESS.COM

CONTENTS

ILLUSTRATIONS

INTRODUCTION

Carol, my wife, and I both felt drawn to homesteading. We both have always had strong notions to utilize whatever means we could manage to take responsibility for our existence, and do so at a local level. With five kids and living for years in rented houses in rural Virginia, that generally meant raising our own food organically, putting up all we could each year, and working our butts off to keep the usual bills paid.

Change didn't seem to come particularly fast, but within a few years our diets changed from the more traditional southern to the nearly vegetarian, our medical ideas shifted from mainstream to herbal/alternative, and our endless sea of rentals ended when we purchased our own land, twenty rough wooded acres on a small river in the Blue Ridge Mountains.

Increasingly we shifted from blindly volunteering our decisions over to certain authorities the way our parents had, to taking back those choices and manifesting in the world what our guts told us was right. We began home schooling the two youngest children which resulted in both being accepted to colleges at sixteen years old, and both making dean's lists. We entered the ranks of the self-employed, achieving success making a living at what we love and experiencing periods full of new challenges. And, we set about building our own homestead ourselves, completely off-grid and as environmentally friendly as possible.

This is the road that led me to the hydraulic ram pump. As the prospects of purchasing land became more real, the understanding of what type of overall alternative energy system we required also became more apparent.

Research at this point turned up ram pumps in a couple of sketchy articles, the main one from the seventies. As I studied the devise, read and re-read the overviews, I couldn't make sense of them. It wasn't until I built a ram that I fully grasped how they function. That understanding, and a much improved design over the original which had been depicted in a VITA (Volunteers In Technical Assistance) article, became the subject of my how-to article, "Poetry in Motion", which ran in the 1996 April/May issue of Mother Earth News. That article led to hundreds of calls and letters which has inspired me to expand on the information from the article and produce this book to better answer all the reader's questions in one handy manual. I hope you will find this book clear and helpful.

The ram articulates the spirit with which we approached our homestead--it is appropriate technology. Rather than drilling into the earth, we've utilized water which is brought to the surface and provided for us already. Rather than installing hundreds of feet of wires and a large electric pump to overcome nature's forces, the ram, with its two moving parts, diverts what is readily available to our own uses.

My greatest hope for this book isn't simply for many who read it to get their own hydraulic ram pumps going, but that it be an example to get each reader discovering many appropriate technologies that can be incorporated into her or his life, wherever that may be. Anyone can homestead anywhere--good choices can be made in the city or in the country. Ultimately, homesteading is taking responsibility for ourselves and focusing our use of resources into wise directions. Let's make efficient use of our available energy and preserve our environment. Let's strive to heal what damage we've already done and bring a halt to unbalanced, unsustainable practices.

PART ONE

The Hydraulic Ram Pump

The hydraulic ram is a non-electric water pump with only two moving parts--a waste valve and a delivery valve. The pump is a prime example of how some homestead utility problems are solved more appropriately and reliably by older, proven technology than mainstream electrified alternatives so fully trusted in today. With only two moving parts to wear out (both easily replaced in the home-built ram discussed in Part III of this book) this pump is the rare technology investment that can last for generations and is virtually maintenance free.

The hydraulic ram was a familiar sight, and sound, in many parts of this country before electrification became widespread. This gadget made its way into my world as our family was closing on the purchase of our homestead here in the Blue Ridge Mountains of Virginia in the mid-1990's. Remaining off-grid was our intention from the start--a choice which carried with it the need to scrutinize and assess the powering of all aspects of the homestead right from the beginning.

Pumping water is a main concern for any home or farm because, short of gravity feeding, it requires a substantial amount of energy. After climbing around our land we knew there were several springs, a large creek, and a river. The houses we had rented previously in the area all were spring fed so we knew digging a well was an unnecessary expense and hassle. None of the springs, however, were uphill from either the house or garden spots. A pumping system from a gathered spring would be necessary.

As we began to plan and purchase our solar electric installation, it became evident that an installation large enough to accommodate pumping the spring water up ninety vertical feet was financially a large burden. Electric pumps suitable for our situation were costly, but worse were the increased costs of more photovoltaic panels, a larger inverter, and an excessive run of electric lines to operate that type of pump off grid. Such a mainstream water system would strap us into debt and we looked hard to find an alternative.

Research turned up a few alternative energy books dating from the 1970's with brief articles on ram pumps. Discussing days past with old timers in town at the weekly bluegrass jamboree at Cockram's General Store revealed a little more information on the subject. Mostly, however, the older crowd focused on the thrilling "CHUCHUNK" that "boomed through the holler" when a "good sized" ram was operating rather than on the operation of the pump itself.

Although a couple of sources discussed the ram in fair detail, a clear explanation of how the machine operated by picture, description, or discussion remained elusive. Perhaps I was just dense, but until I began building my first ram the entire operation of the machine remained a mystery to me.

Understanding (How The Ram Pump Works)

Even though the ram's operation is very simple, it is quite difficult to find a decent description of it. So I am glad to offer the following explanation of this operation in understandable terms (See figure 1-B). The hydraulic ram uses the force of water running downhill through a pipe to then pump some of that water up hill to a site higher than the source.

Water enters the system by running from some source (a spring box or dammed creek for example) into an intake pipe. The water moves downhill through the pipe some distance, enters the pump, and then exits the pump (and therefore splashes out onto the ground) through an open "waste" or "clack" valve. This produces a moving column of water. By allowing the column of water to flow downhill freely exiting through the waste valve, the kinetic energy of the moving water is prepared to work the pump.

Eventually the flow of this water increases to a point that its velocity overcomes the tension of the waste valve spring and it slams closed. By abruptly stopping the flow of the column of water a tremendous pressure is created. This pressure can't very well back out the inlet pipe, rather it pushes forwards through the street bend, through a one way "check" valve, and into an air compression chamber. The compression chamber in the how-to section of this book is simply a length of pipe capped at the top which remains partially full of air. In several cast iron, mass produced rams, the air chamber is a rounded section of metal looking rather like Darth Vader's helmet (Figure 1-A).

The water whooshes past the check valve (below the air chamber) with all its fury squishing the air in the compression chamber until the kinetic energy is converted to, and rests for an instant as, potential energy in the form of compressed air. Next the compressed air pushes the water back out of the chamber with all the energy it has stored. Since the check valve is a one way valve the water being forced backwards cannot return from whence it came; therefore this water is forced through the only place left to go... the exit gate valve and into the delivery pipe thus traveling up hill as far as the energy from the compression chamber will take it. Water continues to move as the ram cycles with the waste and check valves opening and closing and eventually reaches its uphill destination.

When water and air are compressed together some of the air mixes with the water. Because of this, air must be reintroduced to the system constantly or else water leaving the compression chamber will take with it all the air in the pump thus ceasing function all together. To avoid this water-logging problem a small "snifter" hole (or valve) must be present below the check valve so that each cycle brings a gulp of air into the pump which rises into the compression chamber and keeps the machine from water logging.

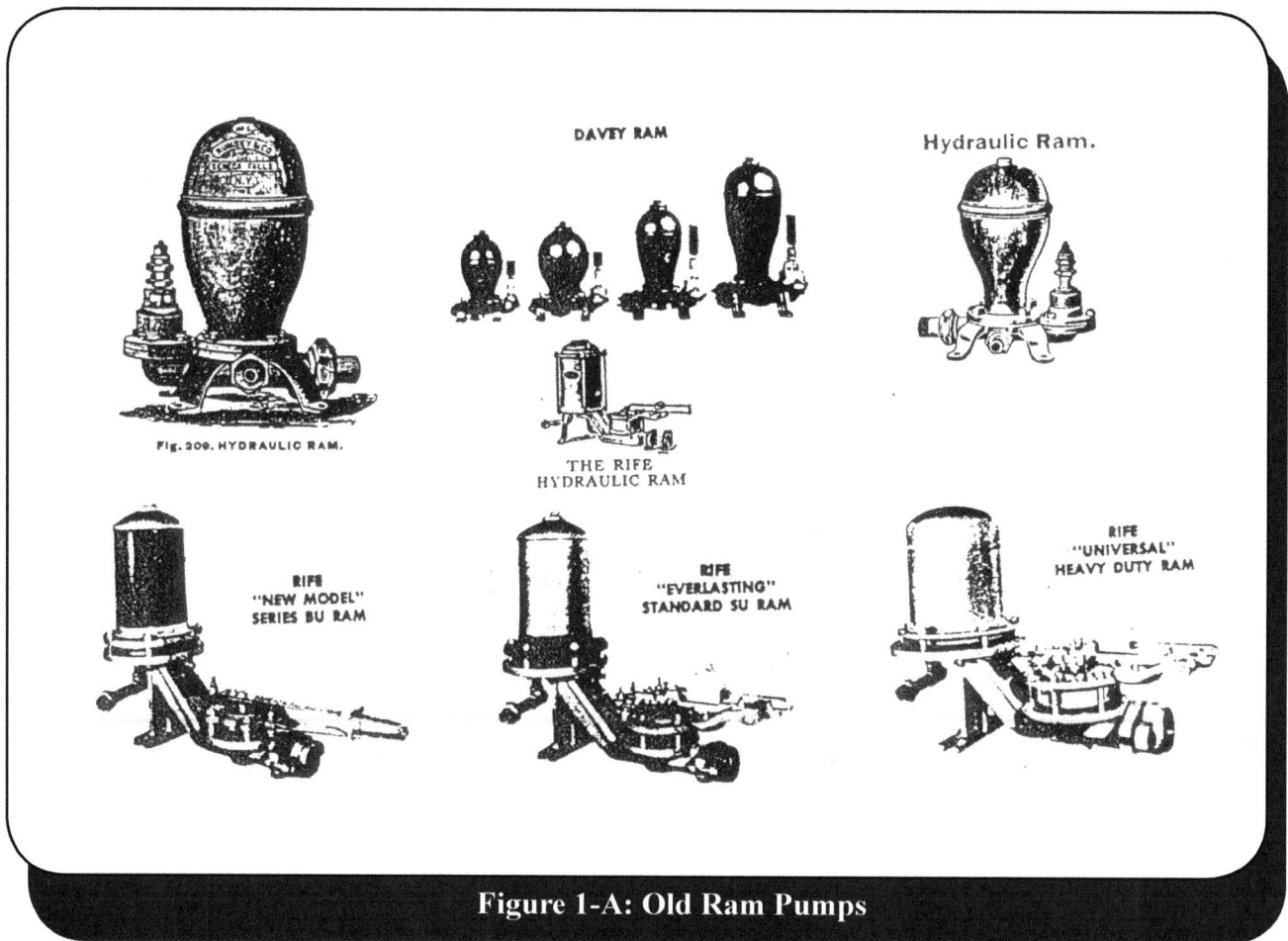

Figure 1-A: Old Ram Pumps

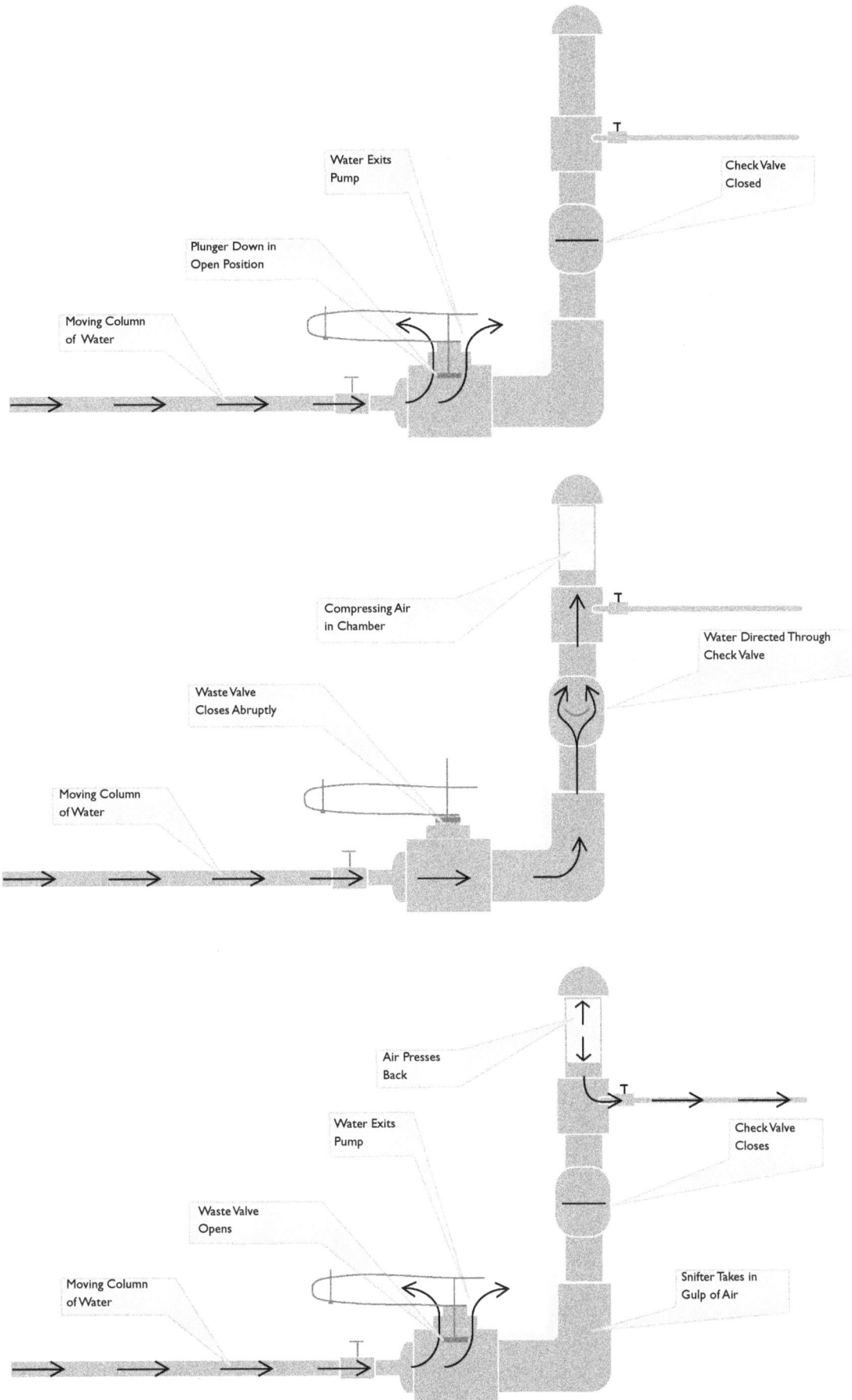

Water Exits Pump

Plunger Down in Open Position

Check Valve Closed

Moving Column of Water

Compressing Air in Chamber

Waste Valve Closes Abruptly

Water Directed Through Check Valve

Moving Column of Water

Air Presses Back

Water Exits Pump

Check Valve Closes

Waste Valve Opens

Moving Column of Water

Snifter Takes in Gulp of Air

Figure 1-B: How The Hydraulic Ram Pump Works

The working of a ram pump is absolutely contingent on the presence of certain geographical conditions. That is, most importantly, you must have a vertical drop from a collection site to the pump site (Figure 1-C).

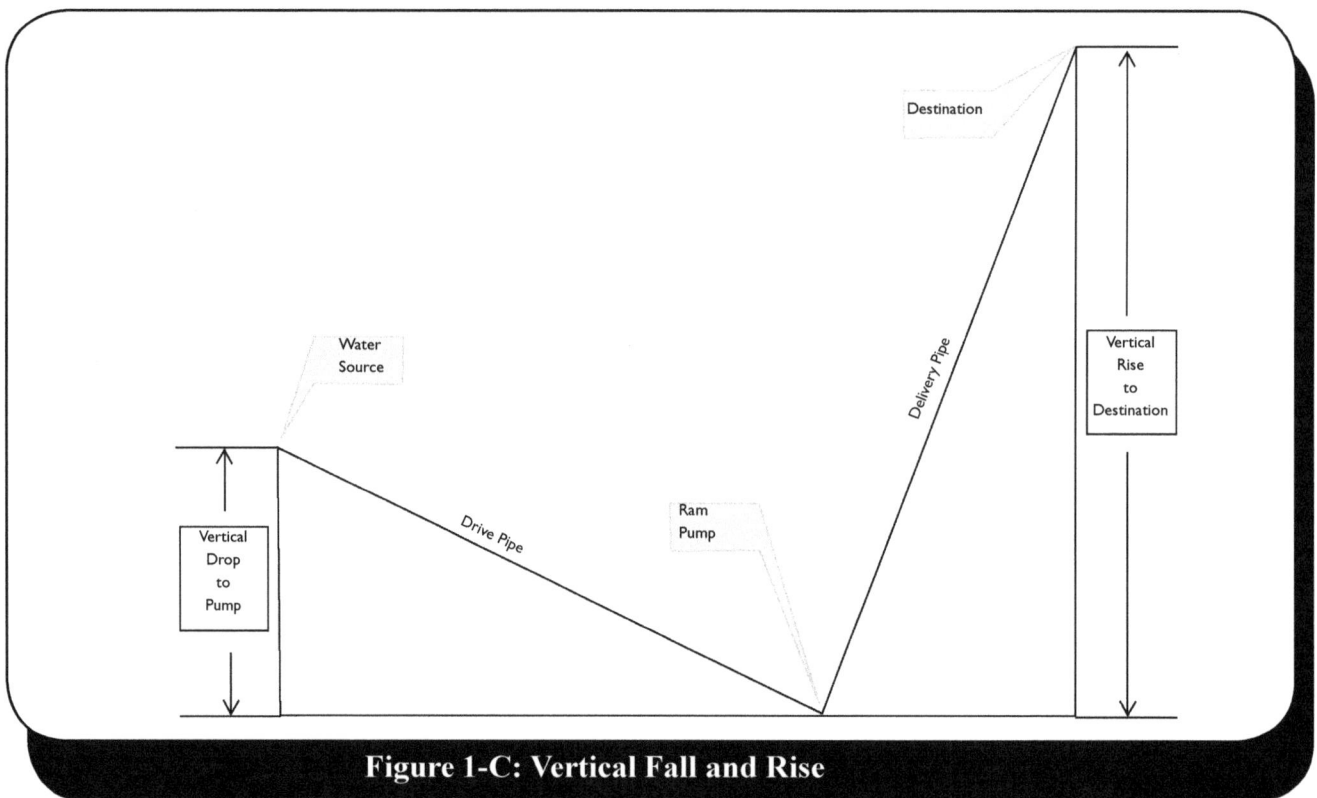

Figure 1-C: Vertical Fall and Rise

Almost any significant flow and fall can be harnessed to run a ram of one size or another. Generally, the minimum vertical drop considered workable is three feet and the minimum volume needed is three gallons per minute. Now, I'm no engineer (I'm a much better builder/tinkerer) and I have run rams on less than three gallons per minute, but it's hard to keep them up and going at slow rates and the water delivered is negligible. Such a situation makes it highly impractical to invest in a pump for these results.

Gauging the flow available in most cases is quite easy. Collect the source into some form of collection box (Figure 1-D) or dam and run it into a single pipe large enough to handle the flow. Then use a container with a known size (a gallon jug or 5 gallon bucket for example) to catch the water exiting the pipe while watching the second hand of the watch to calculate the length of time required to fill the vessel. For example, if I use a gallon milk jug, and it takes fifteen seconds to fill, then there are four gallons per minute available from that source. Repeat this process several times and take an average from all the tries to make sure the results of the test are as accurate as possible.

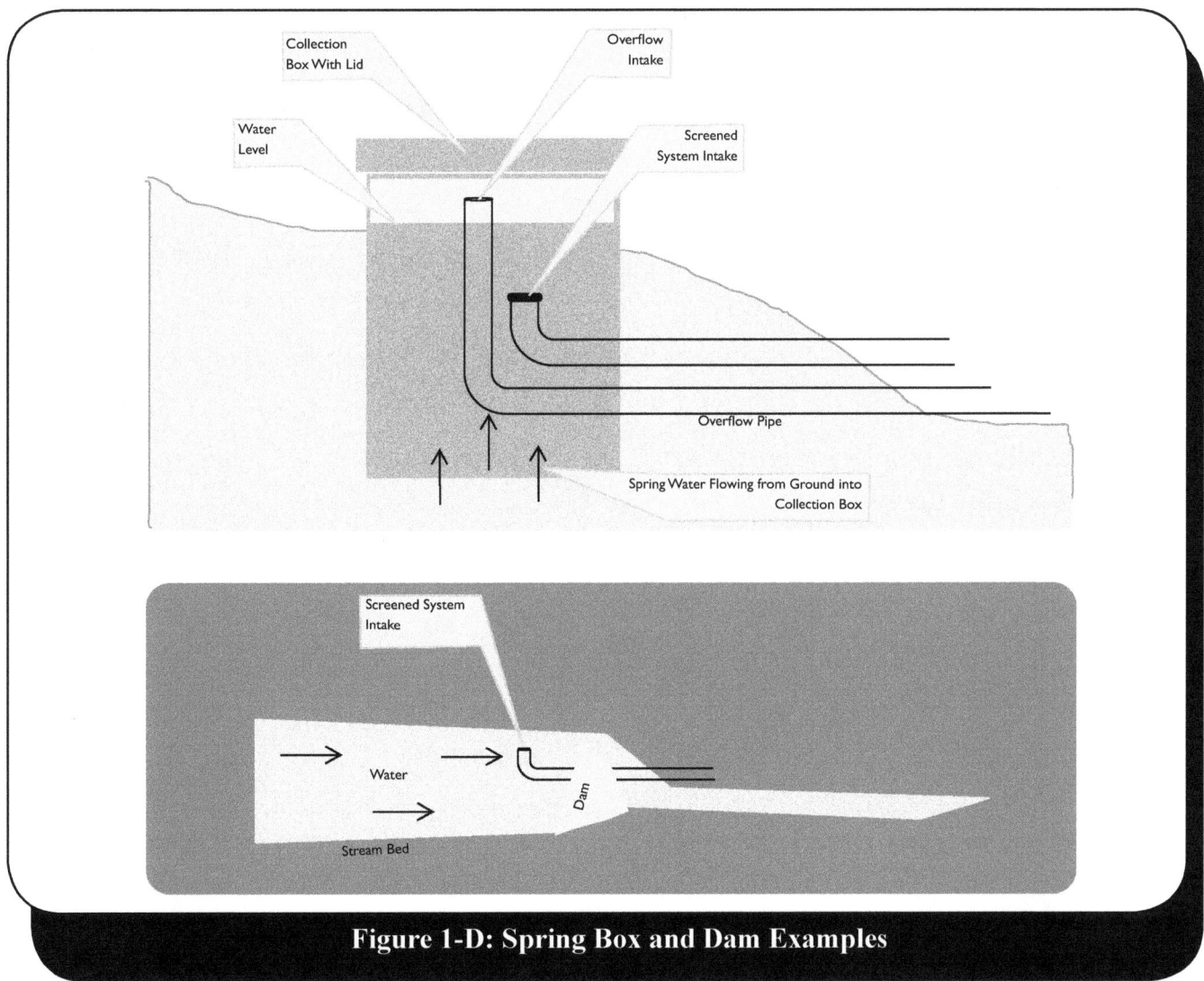

Figure 1-D: Spring Box and Dam Examples

Also, it is imperative to keep in mind that water availabilty can vary hugely different times of the year in different places. Take capacity readings at the driest time of the year if you can be certain this flow is the lowest you are likely to encounter.

The pump plans in this book are for a ram which is adjustable to different available capacities (which has saved me on several occasions). This pump, however, still operates within a certain range, so there is a minimum under which it will not operate--this is also related to the delivery height.

On the other end of the spectrum, if you have an enormous creek or river, gauging available water (if you intend to use large amounts of it) is more difficult to calculate. This book is not designed for such large installations, although rams can be made to pump entire rivers. Once a ram is sized to fittings larger than three inches, it is advisable to use steam fittings and have them welded by a professional. Pressure in such a large pump is extremely high and therefore very dangerous. Check valves must be custom machined from metals capable of such stresses and the rubber in them also must be able to handle such loads. Engineers must be consulted in such instances to determine the size of pipes, dams, etc., for safety reasons. Plus, with such a large pump you would want the huge investment in materials and machinery to be designed to assure safe and long operation. This manual explains the building and installing of a 2 inch hydraulic ram pump constructed from galvanized pipe fittings. While it is possible to change pump size through proportional alterations of all the parts, I have found this 2 inch pump so adjustable as to run from minimal flows, negating the need to size smaller. Also, I believe it is wiser to run parallel 2 inch pumps for larger pumping needs (discussed in detail later) than to up individual pump size.

Figure 1-E: The 5:1 Ratio, Vertical Drop to Drive Pipe Length

Water Storage

Delivery Line Buried Below Frost Line

Ram Pump

Drive Pipe 125'

Spring Collection Box

Vertical Drop 25'

When reviewing your site for ram pump potential, keep in mind this magic ratio for optimizing a ram pump installation. It is 5 to 1 where the drive pipe length is five times longer than the vertical drop from the collection site to the pump site (Figure1-E).

For example, on our homestead here in Virginia the spring collection site is 25 vertical feet above the pump site and the drive pipe is 125 feet long. This vertical drop was ascertained by using a transit which is a small telescope on a tripod with levels and crosshairs which allow the operator to read numbers off a long pole at the downhill end (Figure 1-F). Transits are quite costly, but many hardware stores rent them by the day or a local contractor may be happy to come and take a reading for a small fee.

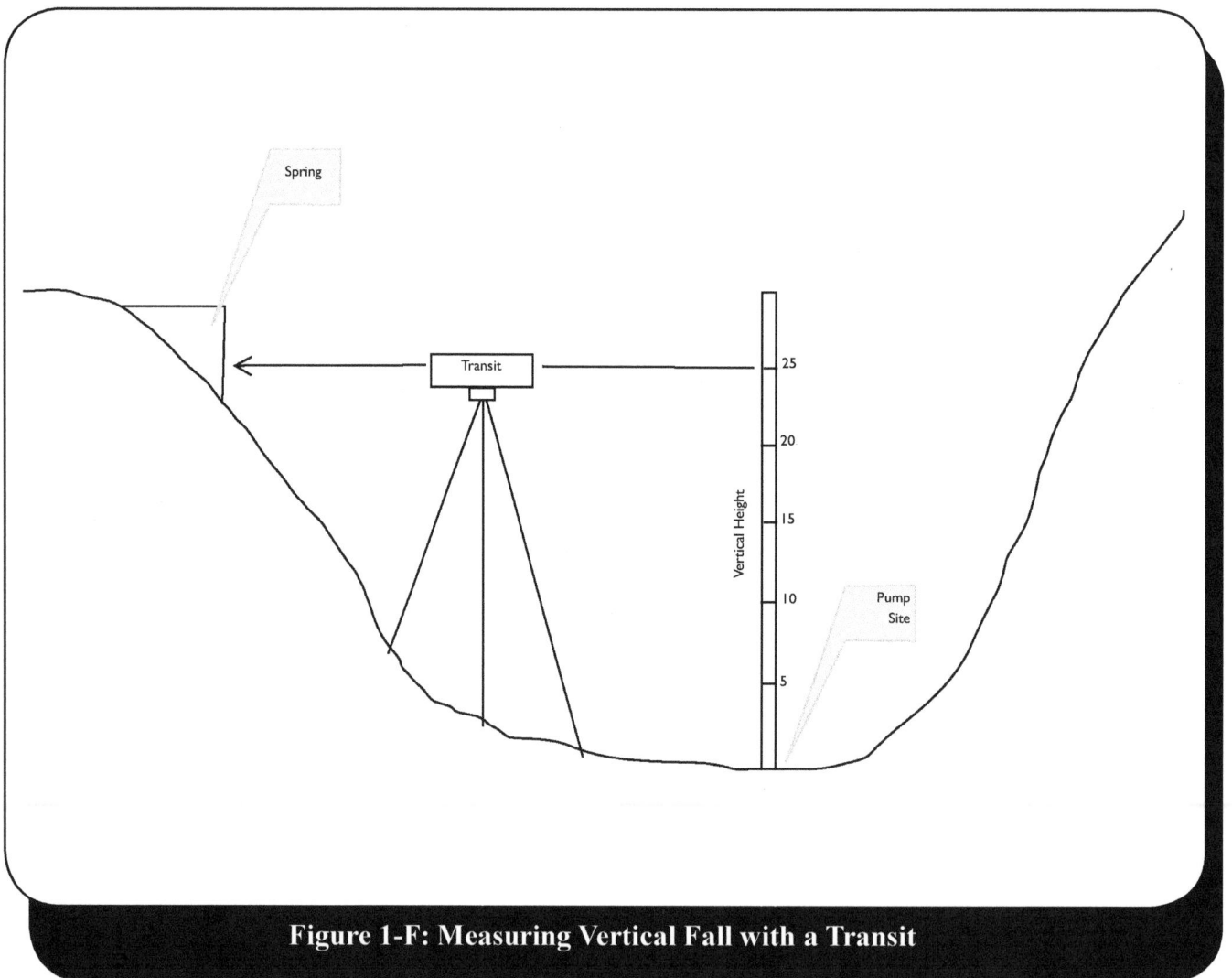

Figure 1-F: Measuring Vertical Fall with a Transit

Another method for figuring vertical distances which works fine for this application requires two people and a four foot level. Starting at the bottom (Figure 1-G) one person should eye across the top edge of the level while the other watches the level bubble. Once it is level, sight cross to a landmark (stone, stump, clump or grass, etc.). That landmark is the same vertical distance from the ground at your feet as your eye is. So, if the measurement from the ground to your eye is 5 feet, then the landmark you sight is also five vertical feet from where you stand. Next, climb up to the landmark, stand on it, and repeat the procedure seeking the next landmark. Repeat this process, adding the vertical feet as you go, until you reach the spot for which you are measuring, and voila, you've got a close measurement of vertical distance.

If you can finagle your collection and/or pump site to achieve the 5:1 ratio then do it. Rams operate at ratios on either side of the 5:1, but they become less and less efficient the farther away from that mark they go until finally they won't run at all.

Also, perhaps a larger consideration than pump function, is the amount of water required at the delivery height--that is, the ultimate destination. An insufficient amount of water at your source, or an impossibly high amount of water required at the destination, can render the ram pump option useless for you.

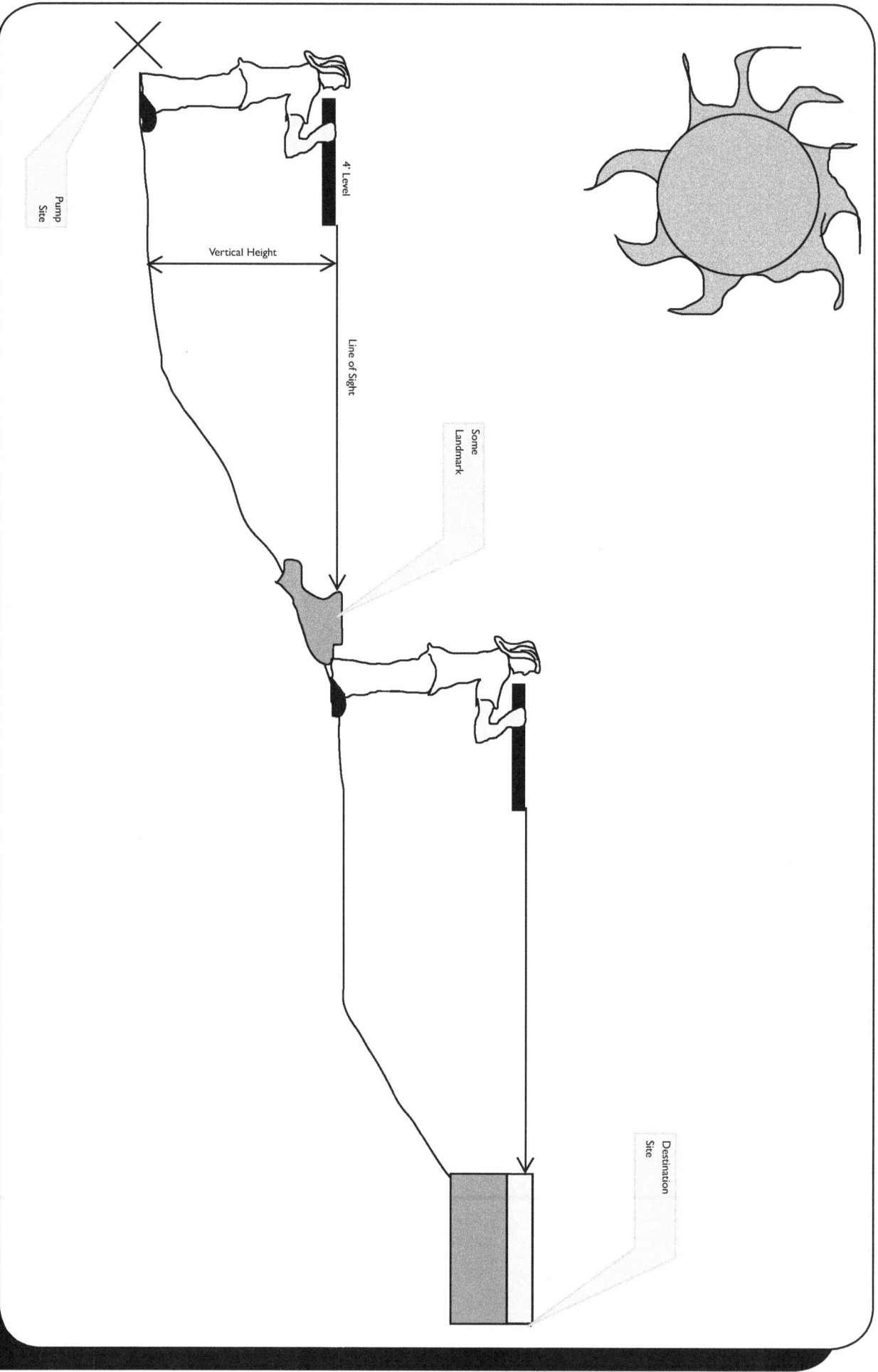

Figure 1-G: Measuring Vertical Height with the 4 foot Level Method

Pump
Site

4' Level

Vertical Height

Line of Sight

Some
Landmark

Destination
Site

A formula for figuring, roughly, what to expect from a ram is:

$$\text{Gallons per hour delivered} = \frac{A \times F \times 40}{H}$$

Where: A=gallons Available from the source

F=vertical Fall in feet

H=vertical Height the water is to be lifted in feet

So, before proceeding any farther, go collect the following information so you can begin figuring how your overall system is going to look and perform:

1) capacity in gallons available at source,

2) volume of water required at the delivery end,

3) vertical drop from source to pump site,

4) drive pipe distance from source to pump site
   (keep drive pipe as straight as possible),

5) vertical rise from pump site to delivery site,

6) delivery pipe distance from pump site to delivery site.

The pump plans in Part III of this book show the exact sizes used for the pump here on this homestead. It operates on 3 to 8 gallons per minute falling 25 vertical feet through 125 feet of 1 inch drive pipe then rising 100 feet through 400 feet of 1/2 inch delivery pipe where it delivers from a trickle to as much as 2 1/2 gallons per minute depending on how well the spring is flowing. 3/8 inch pipe would handle the volume of water on the delivery side of this pump, but 1/2 inch was chosen due to the cheapness and availability of 1/2 inch black plastic pipe. The smaller pipe was not readily available.

Frequently I've found folks surprised at the small flow that comes from a nicely running ram pump. In this world where electric well pumps throw out torrents of water from 1 1/2 inch pipe, watching a gallon a minute come through a 1/2 inch line seems very puny and unexciting by comparison. Rams run continually, so understand that a regular flow of water will be delivered 24 hours a day.

Unlike the customary well and electric pump set-up where the pump brings a massive flow of water directly from the ground on demand, the ram's job is to pump steadily, filling up a cistern of some kind (see a discussion of destination sites in Part II) which is sized according to need. Even when the cistern is full, the ram continues pumping so the overflow from this system must be routed away from the holding tank. This overflow from the ram system can be used in many ways from automatically irrigating the blueberries to filling a swimming pool or livestock trough with fresh water.

With ample water available, the 2 inch pump shown in the plans in Part III of this book provides adequate water to run an average sized homestead. For instance, if the pump delivers 2 gallons per minute (a potential estimate) that's 120 gallons an hour which is 2,880 gallons every 24 hours. That's a lot of water each day being delivered where you need it, for free.

R u n n i n g Rams In Parallel

A good alternative to increasing the size of the ram is to run two or more in parallel. This allows you to combine the pumping efforts of several smaller pumps. Each ram is constructed individually and <u>must have its own drivepipe</u> (Figure 1-H). The delivery pipe may be combined as long as it is large enough to accommodate the volume of water exiting all the pumps simultaneously. The individual drive pipes are necessary because each ram requires the physical situation of a single, uninterrupted column of moving water to create the essential situation to run it.

The multi-pump configuration not only has the benefits over a larger pump of keeping parts smaller, assembly reasonable, and running pressures safer, but, since pumps may be started and stopped separately, a wildly varying amount of water at the source (such as creeks that are bold during Spring thaws but dry up during Fall) can be accommodated which would be impossible to do with a single large pump. Another notable benefit of the multi-ram system is that it allows for one or more pump/s to remain in operation while another is down for repair or maintenance.

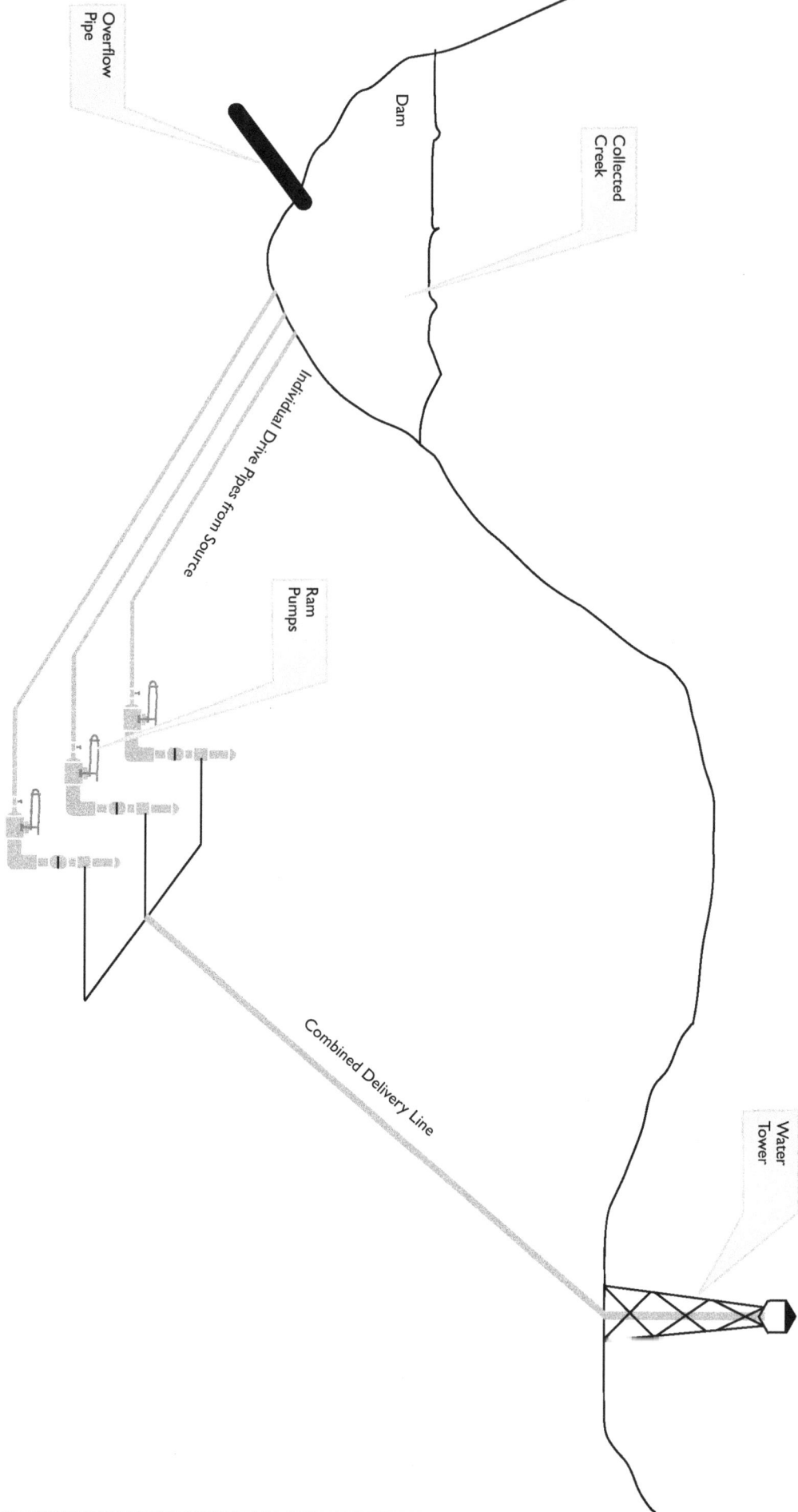

Figure 1-H: Rams Running in Parallel

Overflow Pipe

Dam

Collected Creek

Individual Drive Pipes from Source

Ram Pumps

Combined Delivery Line

Water Tower

As stated earlier, a ram's operation is related to a certain geography. You must have a source of flowing water which then falls at least 3 vertical feet. The first step then is harnessing your flowing water.

Until recently I thought there were only two basic water sources for rams, springs and running ground water (creeks and rivers [and of course ponds which must be fed by either of these]). Then, I went to install a ram here locally and discovered that a well had been drilled, and 1 inch black pipe had been lowered into the casing and a siphon created. Water came out of the pipe down hill from the well at around 7 gallons a minute and ran a 2 inch ram pump of the design explained in this book just fine (Figure 2-A).

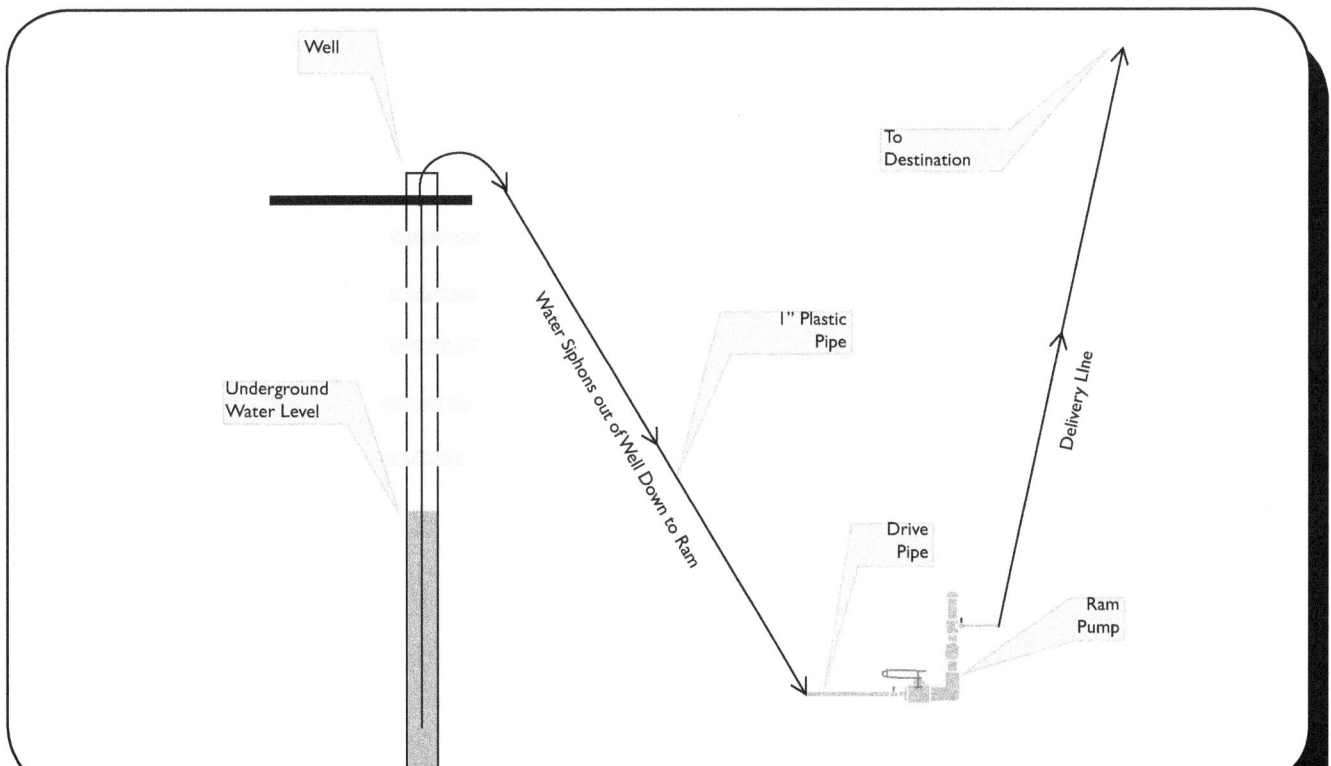

Figure 2-A: Well Siphon System

Here on the homestead we collected a spring. Luckily our spring comes out of a hillside through a trough of rock. We simply laid up a dam using rocks from the site with some fiber reinforced quick-crete. An intake pipe was placed through the bottom of the dam during the building process, and an overflow pipe was layed into a higher part of the dam. It can also be handy to include a drain in the very bottom of a small collection dam like this to get all the water out of the way when working on the dam or intake (Figure 2-B).

Frequently springs are collected by clearing out the area where the water surfaces, either by hand or with the help of machinery like a backhoe, and then setting in place around the source a clean container like a section of large concrete well casing or some other large pipe or box (Figure 1-D). Then, intake and overflow pipes are installed.

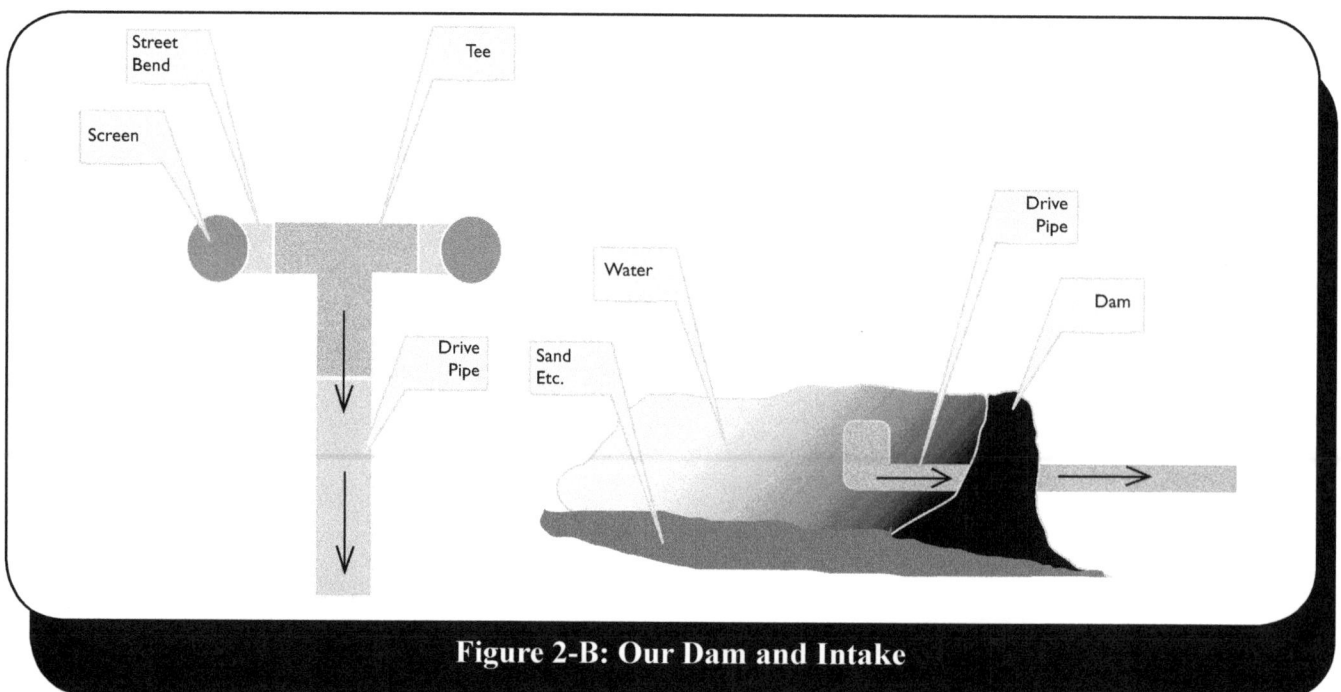

Figure 2-B: Our Dam and Intake

Damming creeks or rivers is a somewhat more challenging task. If there is a way to divert the stream while building the dam in the normal stream bed, it helps tremendously--like cutting into the bank with a backhoe temporarily. Otherwise, you must fight with the flowing water. It's good to note that in many situations you only need to pool the water up a bit to allow the intake access to water which isn't on the bottom with the sand and silt. Simply positioning some rocks or pieces of concrete can many times provide adequate damming (Figure 2-C).

The Intake

The intake is very simple, but is important enough to get its own little section. The point here is to create an intake which minimizes anything other than water from flowing into your system. Frogs are one of the few things that will stop this type of pump in its tracks, and it's a terrible job to clean out a poor critter. Also, sand and particulate matter gunk up filters and sprayers, and then wear on the parts of some pumps and appliances which you may run uphill somewhere in the system.

A very effective yet simple design for an intake is to configure pipe fittings in a manner that keeps the system entry port off the floor of the collection site by orienting openings straight upwards, and covering any opening with screen. My favorite design for this uses a tee and two street bends (Figure 2-D). First, attatch the tee to the end of the drive pipe at the source. Then, screw a street bend into each end of the tee and orient them upwards (place about three wraps of Teflon tape on the male threads of each connection). Finally, cover each intake opening with brass or plastic screen by wrapping the screen over the opening and holding it in place with pipe clamps.

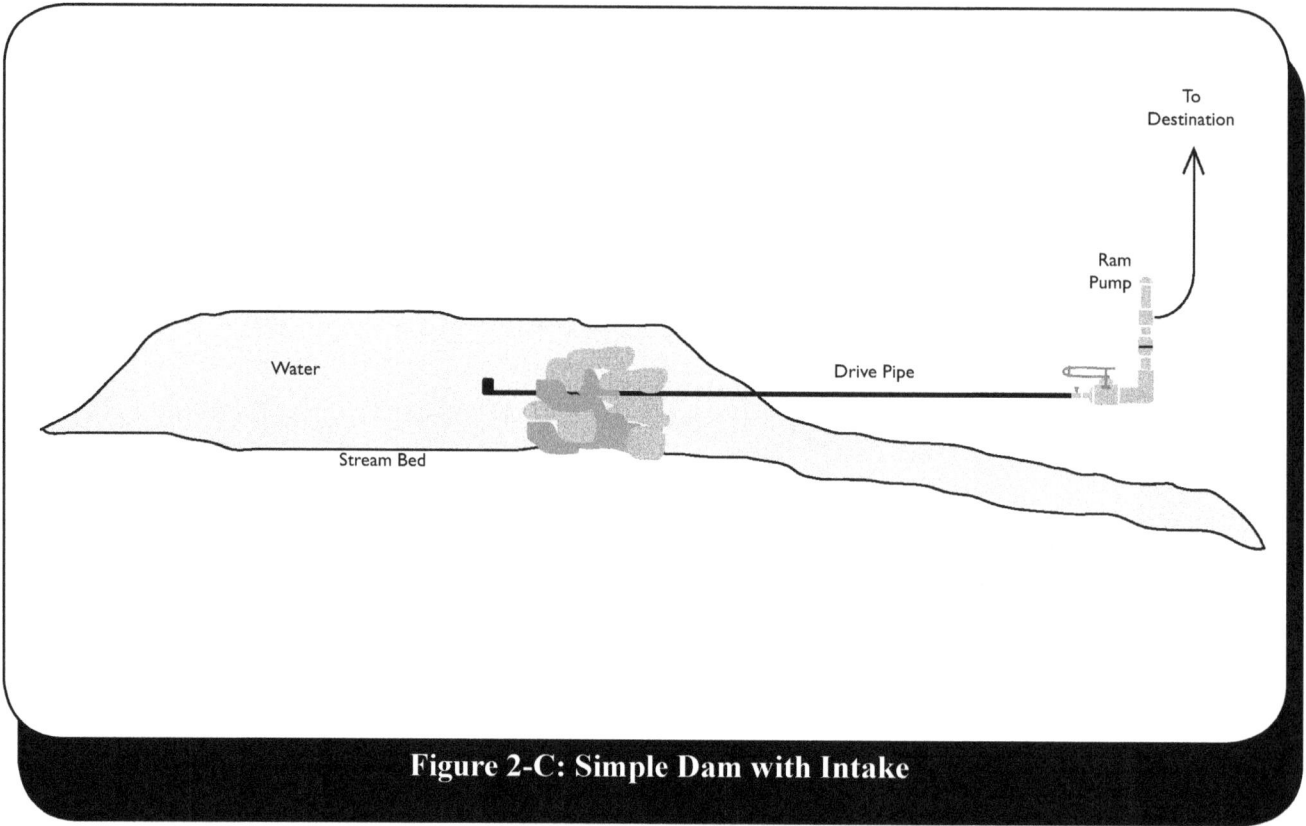

Figure 2-C: Simple Dam with Intake

To Destination

Ram Pump

Drive Pipe

Water

Stream Bed

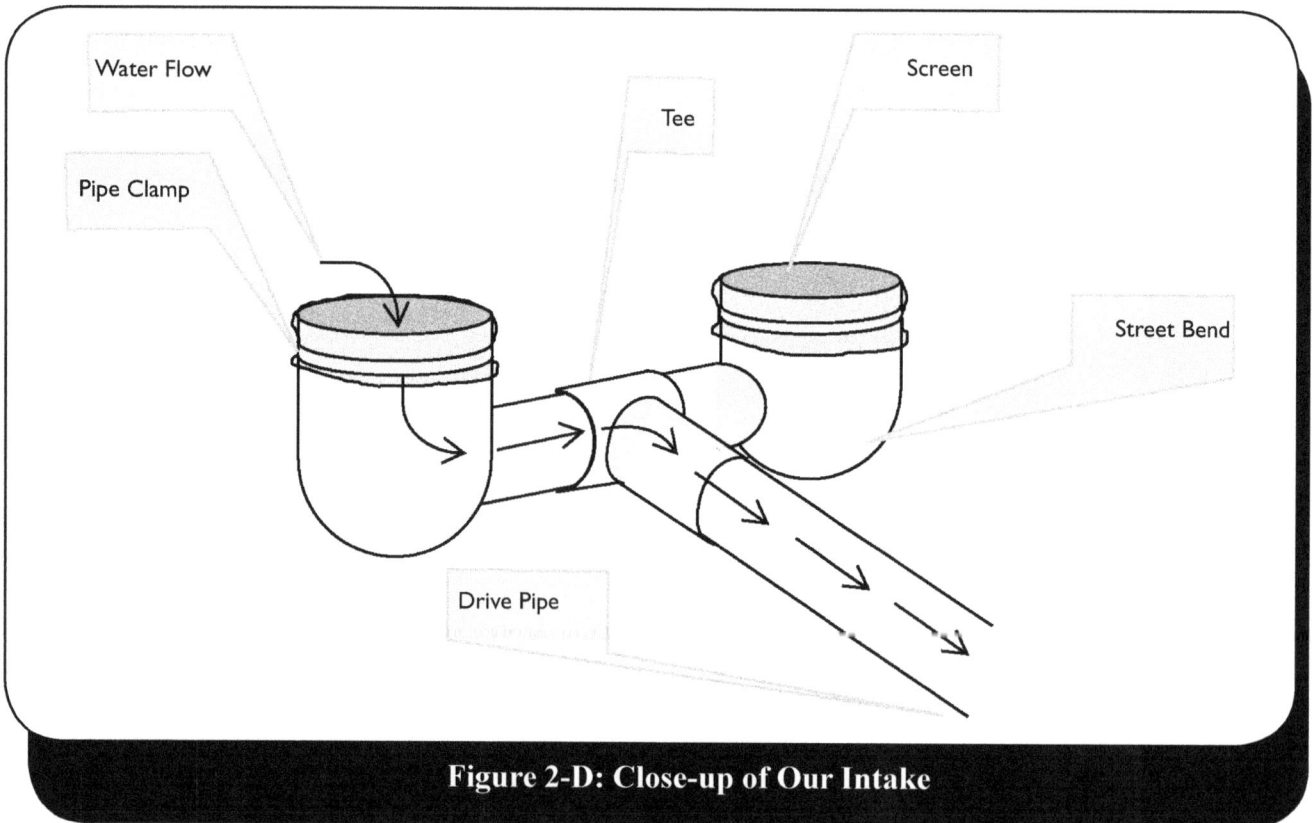

Figure 2-D: Close-up of Our Intake

Water Flow

Screen

Tee

Pipe Clamp

Street Bend

Drive Pipe

Entire System Overview

With your source improved, you can get an accurate idea of what type of flow you have to work with and thereby know for sure a ram will work sufficiently well to continue along the installation process. Once this is established, you need to look at how your entire system will work and go together.

Perhaps the most appropriate ram system is one where a ram pumps water up to an elevation much higher than the ultimate destination to a storage cistern which in turn feeds the system with gravity pressure (Figure 2-E). With gravity-fed water there is no machinery needed to create the pressure for the system, so there isn't maintenance or power requirements.

This, however, generally means going sixty vertical feet or more above the site where the pressurized water is needed, which isn't always practical with a ram pump as the higher the water is pumped, the less water is pumped per minute.

Here at our homestead, the house and gardens are atop a hill. There isn't any land above the destination site to place a cistern. We could have constructed a water tower, but what a pain and expense! Rather, we installed a cistern composed of five plastic 55 gallon drums laid on their sides and connected with pipe, and then used a small 12 volt diaphragm pump to pressurize the 1/2 inch line to the house (See Our System Under System Schematics Appendix B). It provides 45 psi, enough for all the tasks here including a dishwasher, and uses very little current. The pump, a Shur-Flo 2088 series, even has its own built in pressure switch.

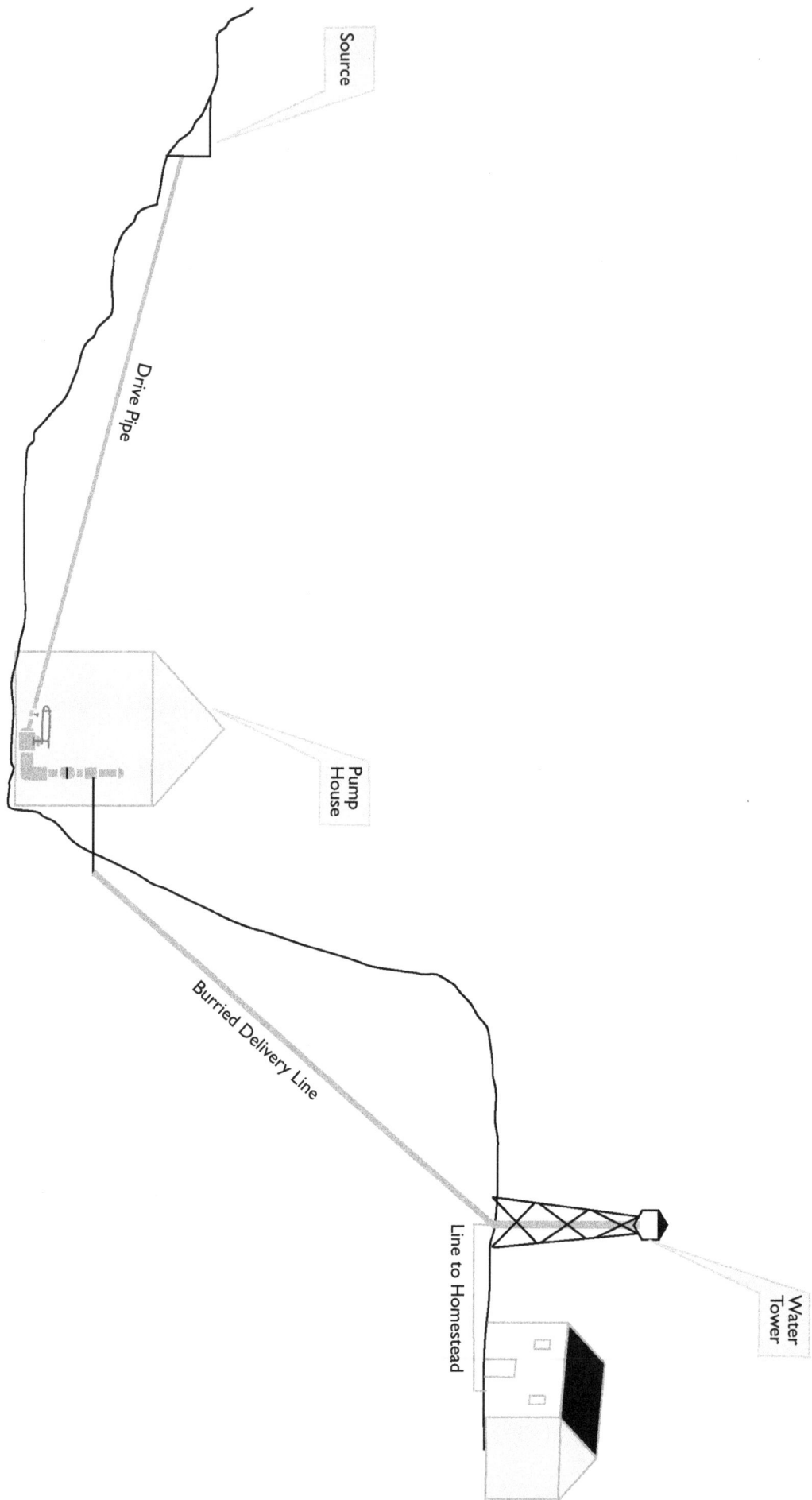

Figure 2-E: Ram System with Gravity Fed Pressure

Source

Drive Pipe

Pump House

Burried Delivery Line

Line to Homestead

Water Tower

Essential to keep in mind too is that your ram pump will run continually. So, at the storage cistern where the water is delivered, that cistern will require an overflow pipe to carry off the extra water when the storage tank is full. We allow the cistern overflow to run into our raised beds thereby automatically irrigating the garden (Figure 2-F). You may wish to use this water for irrigation, garden fountains, feeding a pond, or to simply find a good means to return it back down the hill.

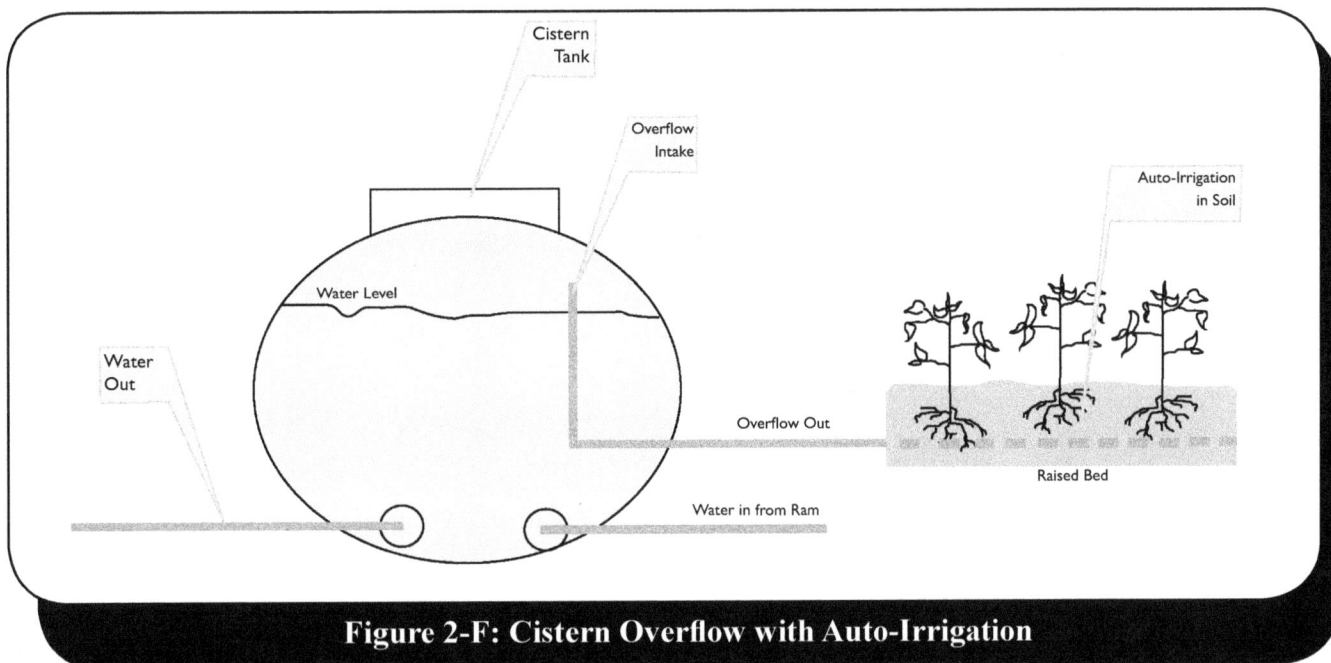

Figure 2-F: Cistern Overflow with Auto-Irrigation

The only other major worry for the overall system at this point is freezing. An hydraulic ram pump has an advantage over electric pumps in this realm because (particularly with spring water) the flowing liquid is warm as it leaves the source and runs through the system. Since rams run continually, that means warm water constantly flows throughout the pipes and pump thereby keeping it all warm to a certain extent.

Now, here in Virginia, I can get away with the pump setting out on the ground running out in the woods nearly all Winter long. Much more common than the pump freezing up is the small 1/2 inch delivery line freezing somewhere along the hillside. Of course, if the pump should malfunction and cease to operate, it will freeze up and parts may break. Therefore, I suggest taking regular plumbing precautions for your area to avoid freezing- -i.e. burying lines below the frost line and constructing some type of pump house to help maintain a higher temperature around the ram.

Efficiency

Especially if you have marginal conditions to run a ram pump, efficiency will be of critical importance. Of paramount concern is keeping the drive pipe as straight as possible. If this isn't feasible, at the very least try to get the last 1/3 of the drive pipe in front of the ram perfectly straight. It is best to run the drive line in galvanized pipe. If the cost of metal pipe is prohibitive, schedule 40 PVC will work okay for the first 2/3 of the run. I have seen 1 inch black pipe rated for a high psi used to drive one of these rams as well, but it was very difficult to keep it straight.

The other major factor to maximize hydraulic ram efficiency is getting as close as possible to the aforementioned 5:1 drive pipe length to vertical drop ratio.

One other trick which may be helpful to your situation is the inclusion of a vent pipe. There are conflicting views around about the use of vent pipes in the drive pipes of hydraulic rams, but here's the low down I got from an engineering type: when adding a vent pipe to the drive line of a hydraulic ram pump, you must begin your drive pipe figures from the vent pipe as if it were the source. For example, if you have a very difficult scenario where the spring you collected is high up a mountain side, but nowhere close to a decent pump site, you could use a lengthy piece of pipe to move the water around the mountain side to where a 5:1 ratio to a good pump site is possible. There, place a vent pipe which is tall enough to reach the vertical height of the source, at a tee in the drive pipe, and there shoot 5:1 as straight as possible to the ram (Figure 2-G).

The experiments with these vent pipes I've witnessed seem to make the ram stroke faster; that is, the plunger in the waste valve recovers to the open position much quicker than without the vent, seeming to have a positive affect. But, at the same time, less (and even no) water was delivered to the destination site even when the pump ran very well. I think the pressure available for ramming is reduced when using the vent pipe if it isn't positioned along the drive pipe to achieve the 5:1 ratio.

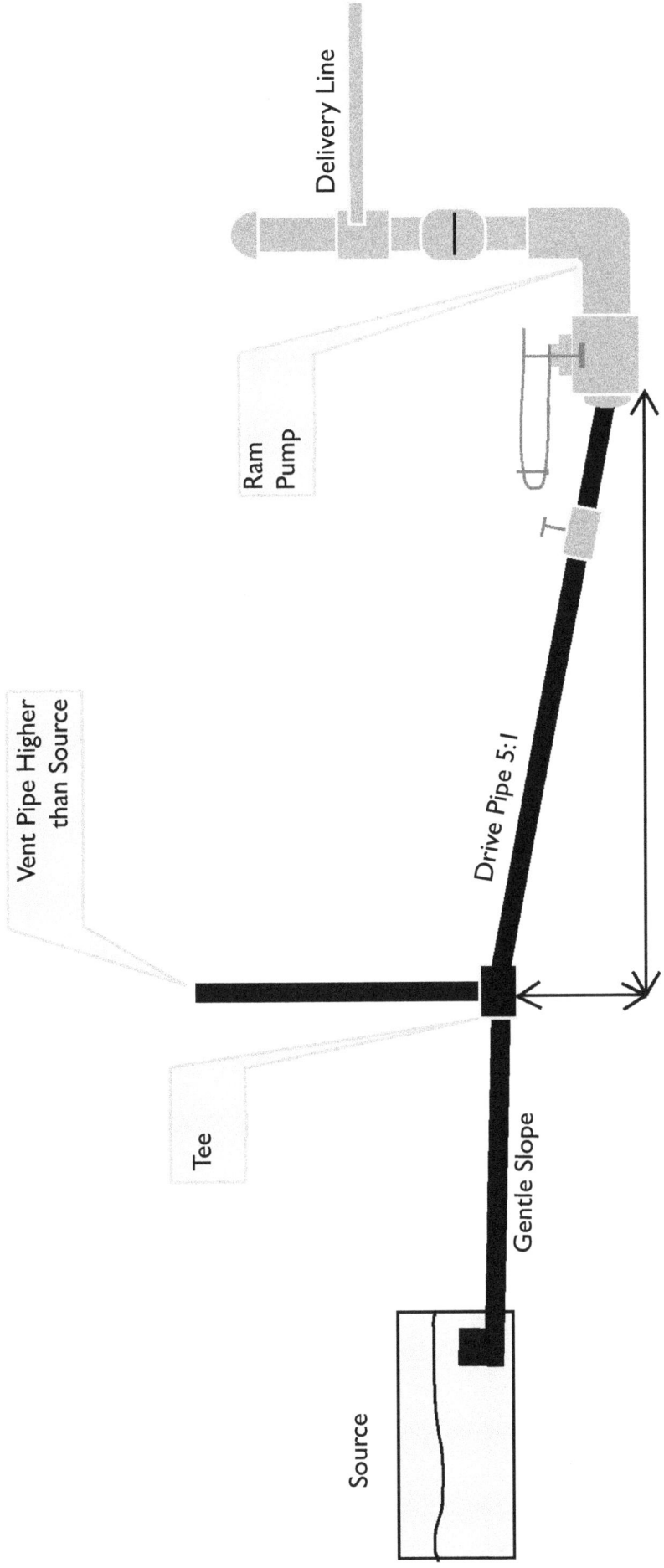

Figure 2-G: Drive Line with Vent Pipe

Delivery Line

Ram Pump

Vent Pipe Higher than Source

Tee

Drive Pipe 5:1

Gentle Slope

Source

PART THREE

Building An Inexpensive
Ram Pump

This ram pump can be built by the average person in a remarkably short time for around $150 depending on costs for plumbing supplies in your area and what kind of usable junk you keep around your place.

To build this average sized ram (Figure 3-A) you will need:

Part	Quantity
1" union	1
1" gate valve (optional)	1
1" nipple (only w/gate valve)	1
2"X 1" bushing	2
2" tee	1
2" street bend	1
2" check valve	1
2" nipple	2
2" X 2" X 1/2" T	1
2" threaded pipe 2' long	1
2" cap	1
1/2" nipple	1
1/2" gate valve (optional)	1
40" of 1 1/2" X 1/4" strap	1
3/8" X 4 1/2" bolt w/nut	1
1/4" X 3" bolt w/2 nuts	1
3/8" X 3" bolt w/2 nuts	1
1/2" copper pipe 3 3/4" long	1
1/2" X 1" bolt w/1 nut	2
1/2" lock washer	2
3/8" flat washer	2
1/2" flat washer	1
2" small gauge copper or brass wire	1
roll Teflon tape	1

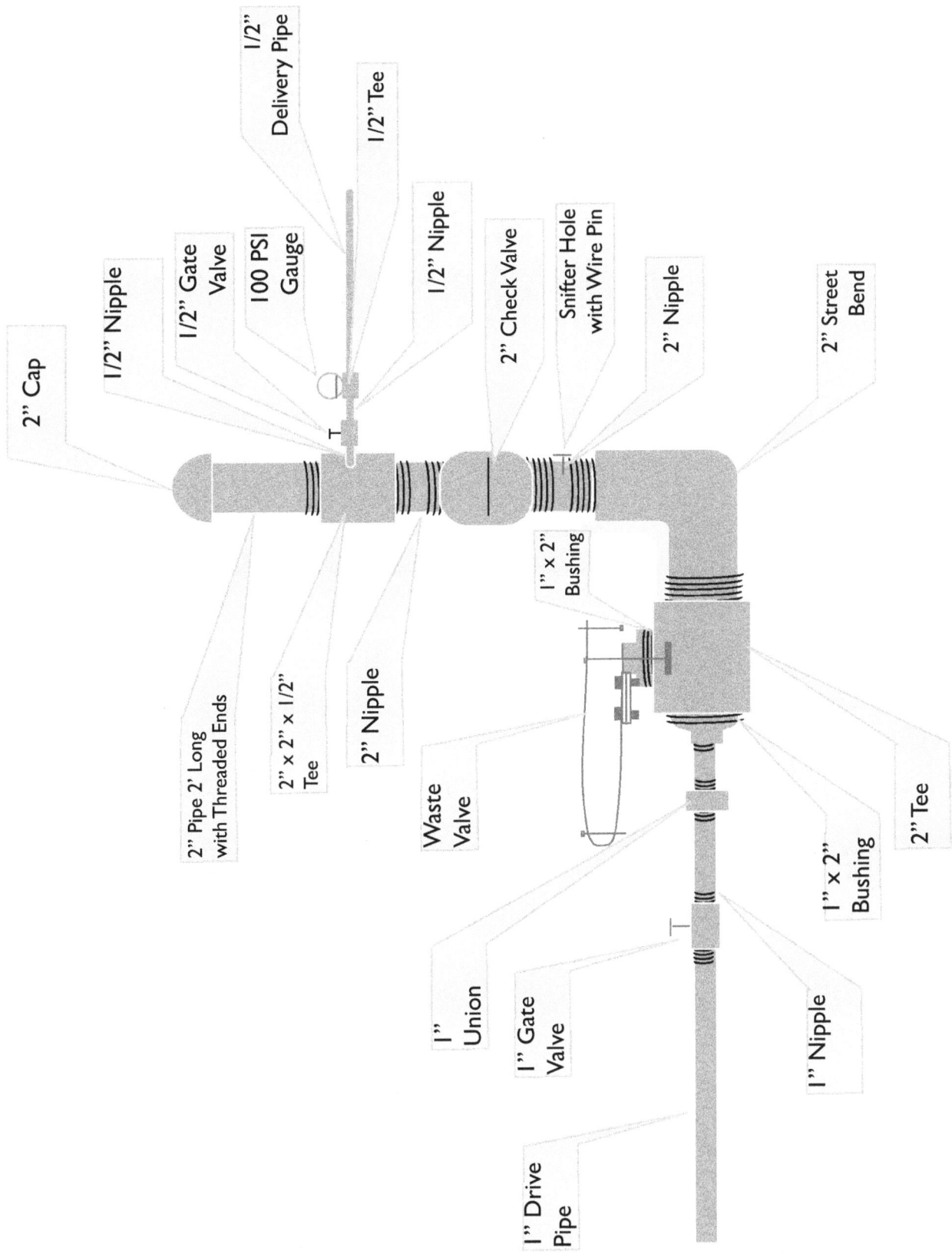

2" Cap

1/2" Nipple

1/2" Gate Valve

100 PSI Gauge

1/2" Delivery Pipe

1/2" Tee

1/2" Nipple

2" Check Valve

Snifter Hole with Wire Pin

2" Nipple

2" Street Bend

2" Pipe 2' Long with Threaded Ends

2" x 2" x 1/2" Tee

2" Nipple

Waste Valve

1" x 2" Bushing

1" Union

1" Gate Valve

1" Nipple

1" x 2" Bushing

2" Tee

1" Drive Pipe

Figure 3-A: Home Built Ram Pump

This list of materials should be available at any well stocked plumbing supply, hardware store, or farm supply. When you purchase the check valve, do not buy one which is brass on brass. Water from a spring, creek, or river will carry fine grit which chips away at the soft brass and will wear the check valve. It is better (and cheaper for that matter) to use a check valve (also called a foot valve in jet pump uses) which has a rubber or neoprene type diaphragm inside.

It's REALLY nice to have access to a heavy bench, vise, and several different sized pipe wrenches when assembling this pump--it will make the process a joy rather than an awkward, difficult battle. Most of the construction is simply screwing pipe fittings together, but the waste valve (also called a clack valve by some) requires a bit of metal shop work

W a s t e Valve

The waste valve (Figure 3-B) is a modification of a very dependable design promoted by VITA (Volunteers in Technical Assistance). This is the most complicated part of this ram pump, and if you would like to simply purchase this whole waste valve assembly and just screw it into place on your pump, the local machine shop, Huffman Tool Company, who makes them for me will be happy to send one to you. He makes them only from stainless steel, however, which ups the price from galvanized fittings, but the quality is unbeatable. Contact him at (540) 745-3359 or hufftool@aol.com.

The first step to making the valve is to take one of the 2" x 1" bushings, and to machine the inside surface where the rubber washer/seal on the plunger will bear against it when in the closed position (Figure 3-C). This process can be done in one of two ways. The best is to chuck the bushing into a metal lathe and then have a sharp bit, which is held by another arm of the lathe, enter into the bushing where the uneven casts surface is machined away to a nice, perfectly flat surface. The machinist here in town does mine for $10 which is probably a reasonable average to expect most anywhere.

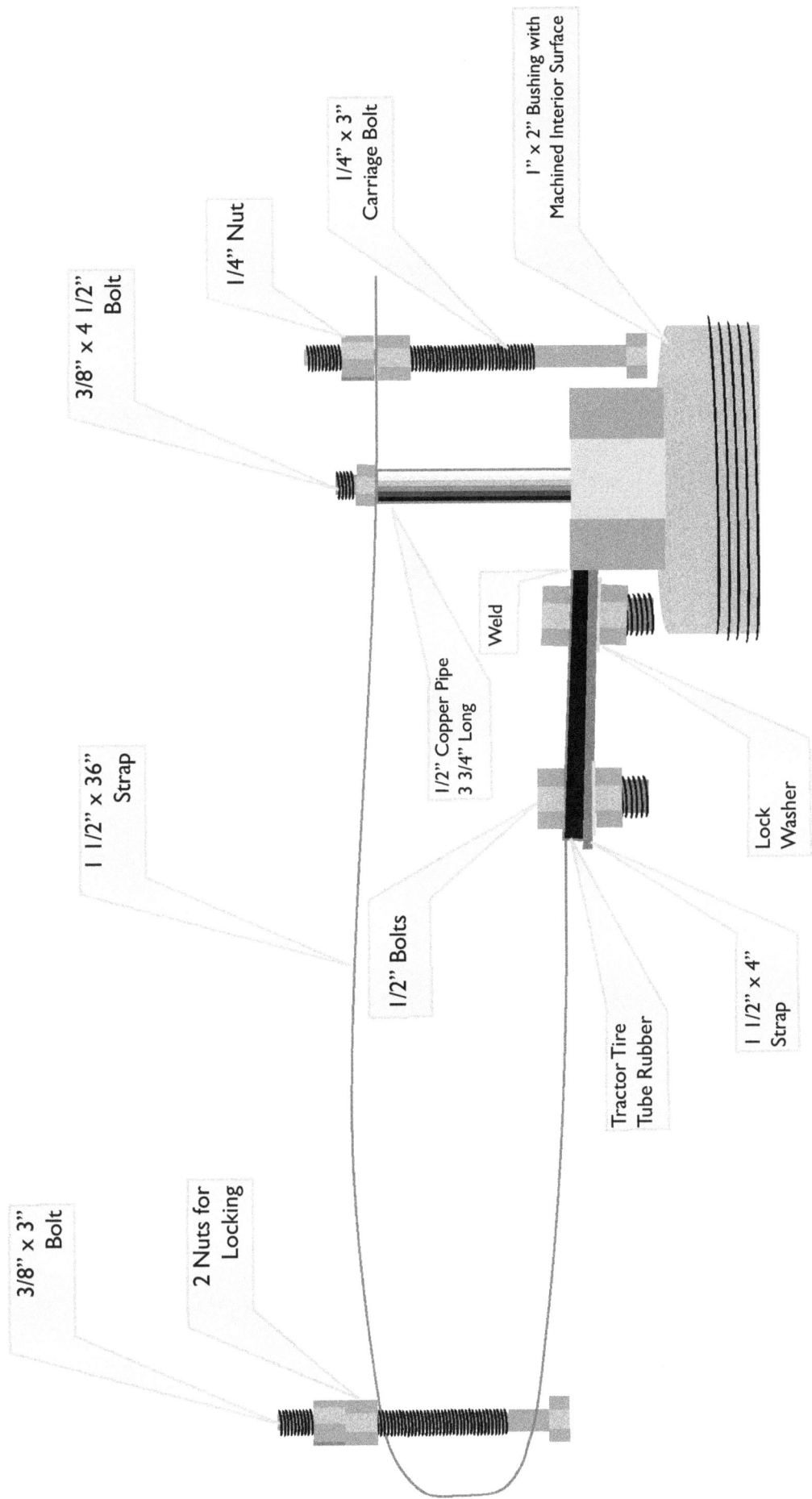

Figure 3-B: Waste Valve Assembly

- 1/4" x 3" Carriage Bolt
- 1" x 2" Bushing with Machined Interior Surface
- 1/4" Nut
- 3/8" x 4 1/2" Bolt
- Weld
- 1/2" Copper Pipe 3 3/4" Long
- 1 1/2" x 36" Strap
- Lock Washer
- 1/2" Bolts
- Tractor Tire Tube Rubber
- 1 1/2" x 4" Strap
- 3/8" x 3" Bolt
- 2 Nuts for Locking

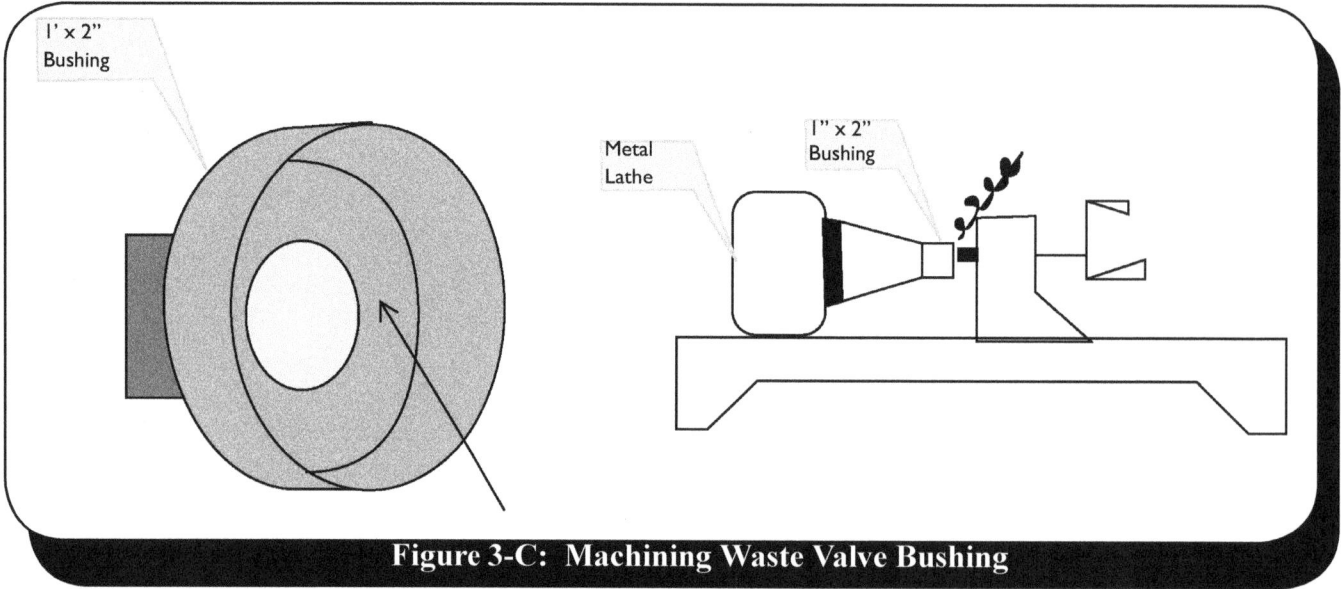

Figure 3-C: Machining Waste Valve Bushing

1' × 2" Bushing

Metal Lathe

1" × 2" Bushing

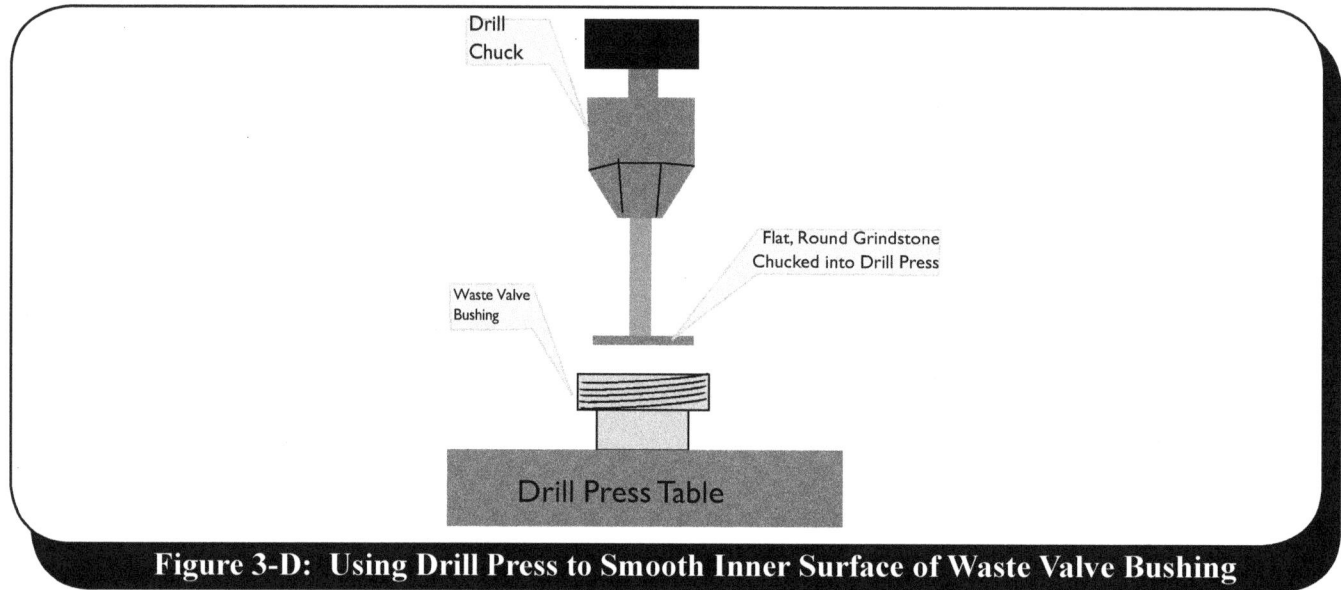

Figure 3-D: Using Drill Press to Smooth Inner Surface of Waste Valve Bushing

Drill Chuck

Flat, Round Grindstone Chucked into Drill Press

Waste Valve Bushing

Drill Press Table

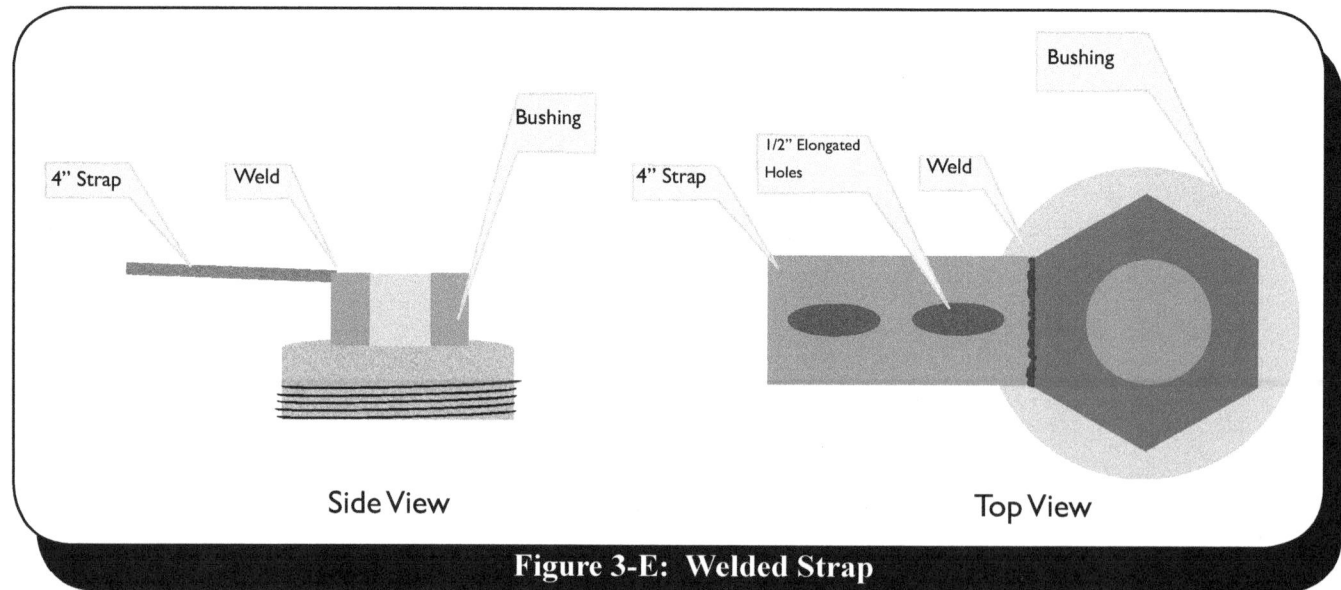

4" Strap

Weld

Bushing

Side View

Bushing

1/2" Elongated Holes

4" Strap

Weld

Top View

Figure 3-E: Welded Strap

The other method is to use a flat grinding stone bit chucked in a drill press. Clamp the bushing upside down on the press table and then lower the spinning stone into the bushing and grind the rough cast surface until smooth (Figure 3-D).

While at the shop, for those who don't have the facilities and experience to cut and weld metal, have 4" cut off the 40" piece of 1 1/2" X 1/4" strap and have it brazed or welded in place on the top of the bushing close to, but not covering, the 1" threaded hole (Figure 3-E). The galvanizing needs to be ground off the bushing to allow the brazing or welding hold properly.

Drill two 1/2" holes through the small strap on the bushing which correspond to two of the same size at one end of the 36" of strap still remaining so the two may be bolted together with a piece of rubber between them. The machinist does this drilling for me as well because he has an endmill which not only makes this drilling precise and simple, but I have him elongate both holes half an inch long ways in the small welded piece of strap which allows for a slight adjustment when centering the plunger part of the waste valve (Figure 3-F).

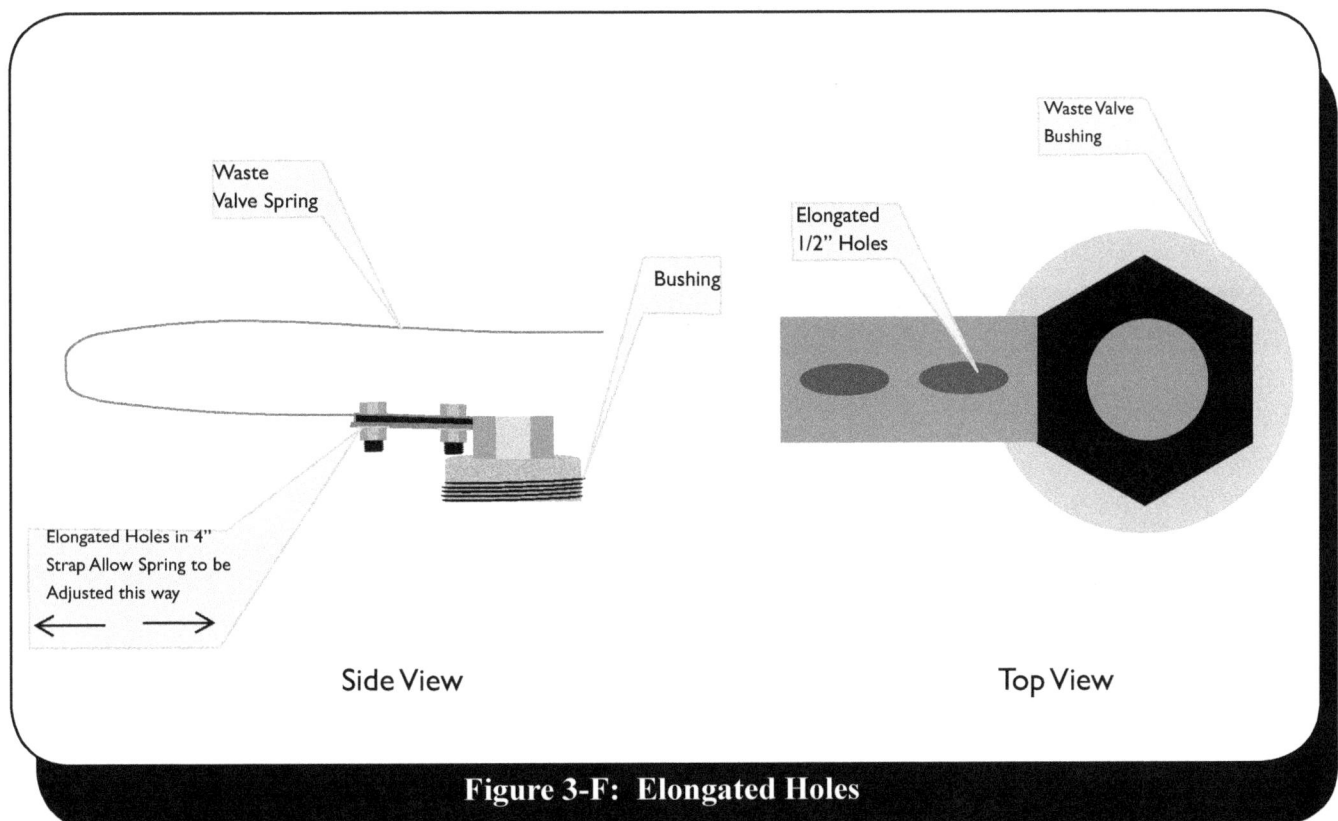

Figure 3-F: Elongated Holes

Next, make a mark on the 36" strap, 16" from the end with the two 1/2" holes and bend the strap around a 1 1/2" pipe centering the mark in the bend to make the waste valve spring (Figure 3-G). Drill two 3/8" holes corresponding to one another top and bottom of the spring right where it flattens out after the half circle bend to allow for the 3/8" x 3" bolt to pass through both holes. Add two nuts, one to adjust the tension of the spring and the other to act as a lock nut to keep the adjustment from moving during operation.

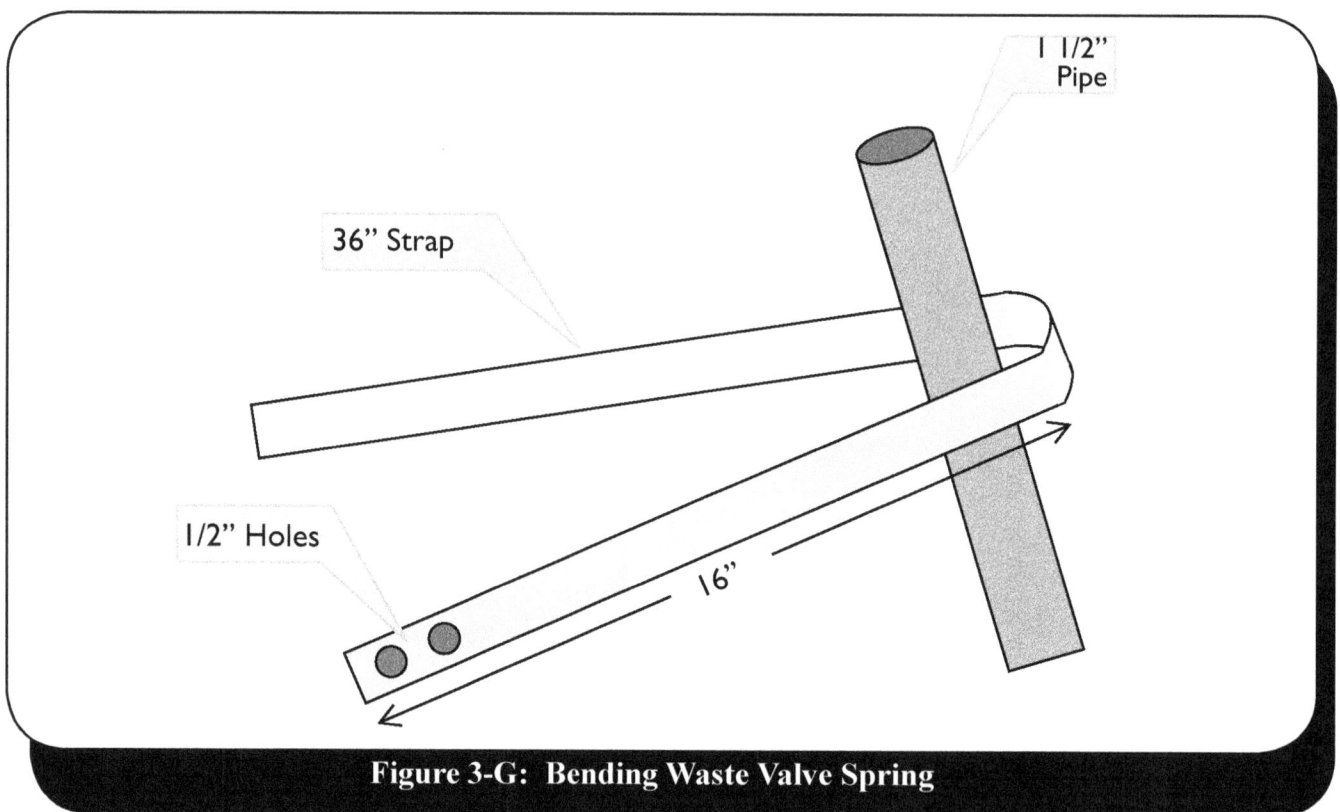

Figure 3-G: Bending Waste Valve Spring

Two more holes must be drilled and then the hard part is over. Bolt the spring in place to the bushing with the two 1/2" bolts finger tight, centered in the elongated adjustment bolt holes. Make a mark in the center of the strap exactly where it passes over the center of the 1" hole in the bushing. Make another mark beyond the first towards the end of the spring directly over the far edge of the bushing (Figure 3-H). Now unbolt the spring from the bushing and drill a 1/4" hole at the mark towards the end of the spring and a 3/8" hole at the mark made directly over the bushing's opening.

Mark Here for Center of
1/4" Hole for Carriage Bolt
Stop

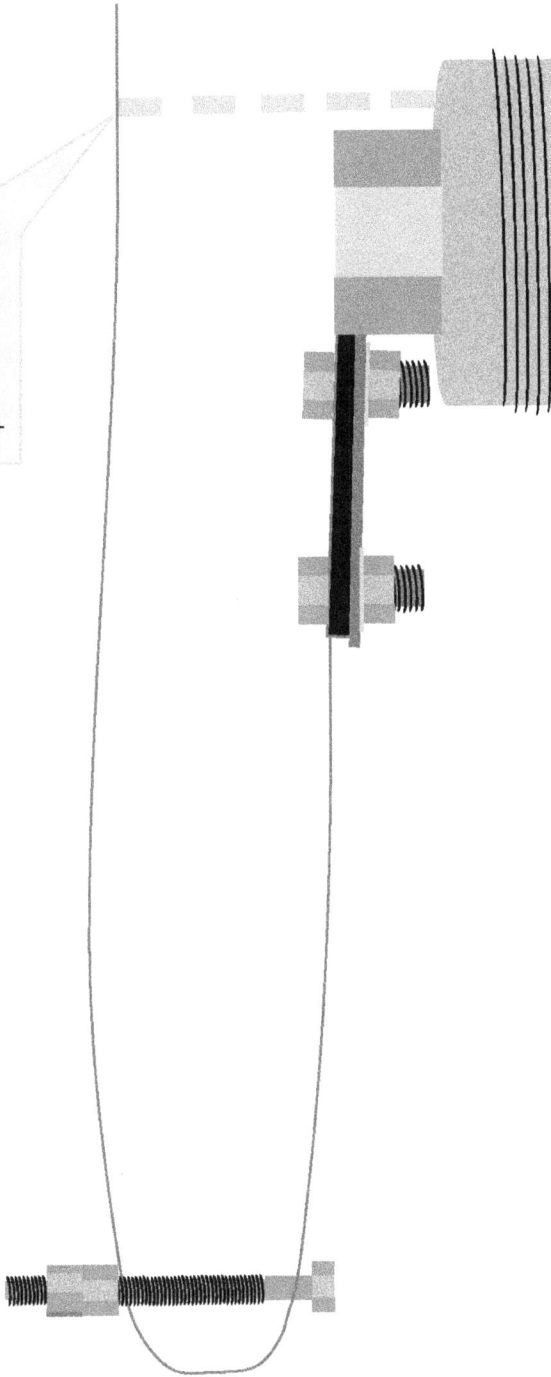

Mark here for Center
of Plunger Bolt Here

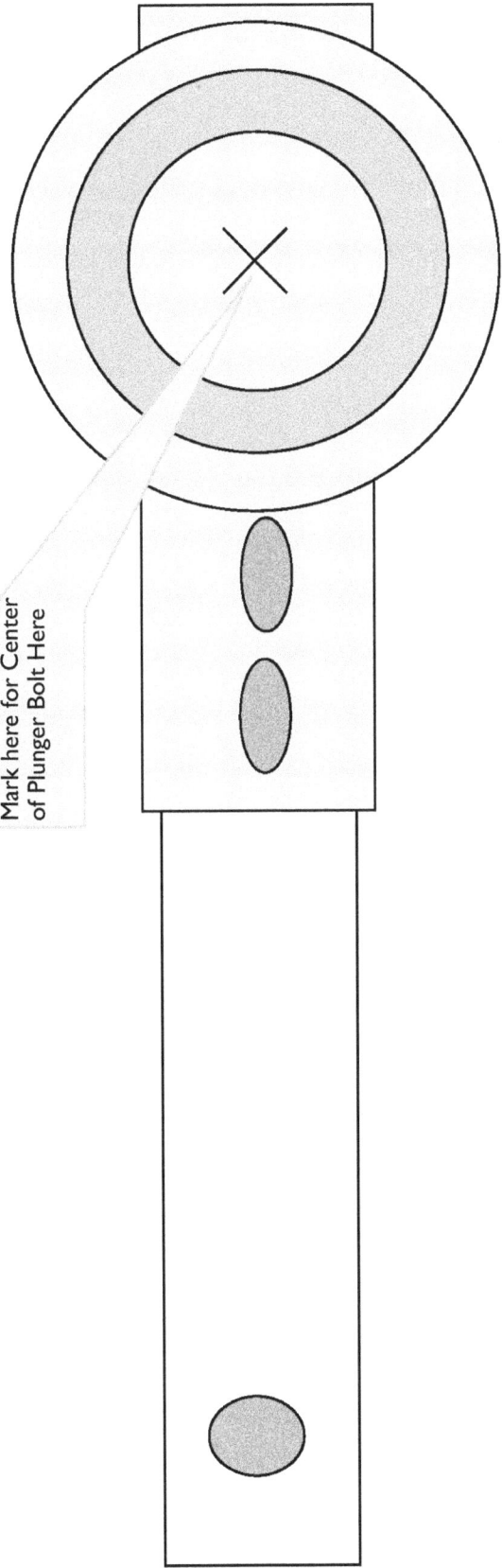

Figure 3-H: Marking for Plunger and Stop Holes

To complete the waste valve, cut a piece of rubber large enough to sandwich between the waste valve spring and the stud it bolts to on the bushing. Used tractor tire tube works well for this. Cut holes for the two 1/2" bolts to pass through and then bolt the spring tightly in place using a lockwasher under each nut.

Take the 3/8" bolt and place on it a 3/8" flatwasher, then the 1/2" flatwasher, then a piece of rubber cut into a 1 3/8" outer diameter circle, and finally the length of copper pipe. Pass the fully loaded bolt through the machined bushing and then through the 3/8" hole in the spring. On the top side of the spring add a lockwasher and a nut and tighten snugly the whole works (see figure 3-I).

I've experimented with several different types of rubber for the waste valve and the best for this abusive spot is soft radial tire sidewall. It can be cut with a sharp knife or leather scissors. Orient the soft outer side upwards so that it presses against the machined edge inside the bushing making the waste valve seal. This rubber makes a great seal and has nylon cord woven into it for durability. If your waste valve is properly aligned during operation so that it doesn't rub on the side of the bushing, this rubber clack should last for several years.

It is imperative that this valve open and close in perfect alignment so that the rubber washer closes completely on the machined surface inside the bushing and that it not bump or rub the side of the bushing in any way as it opens and closes. Slightly bending the spring or bolt and/or adjusting the two pieces of spring at the elongated 1/2" bolt holes works to tweak out a valve not perfectly aligned.

Finally add the 1/4" bolt through the final hole left in the spring to make the opening depth of the waste valve adjustable. Be sure to place a nut on the bolt before sticking it through the hole, and then one after so that the carriage bolt may be adjusted up or down and then tightened in place. You may add a piece of tractor tube rubber over the head of the bolt by cutting two holes and pulling it down the bolt and over the head to reduce shock to the spring, wear, and noise, but it is completely optional.

The waste valve is now complete and needs only to be screwed into the system at a later point.

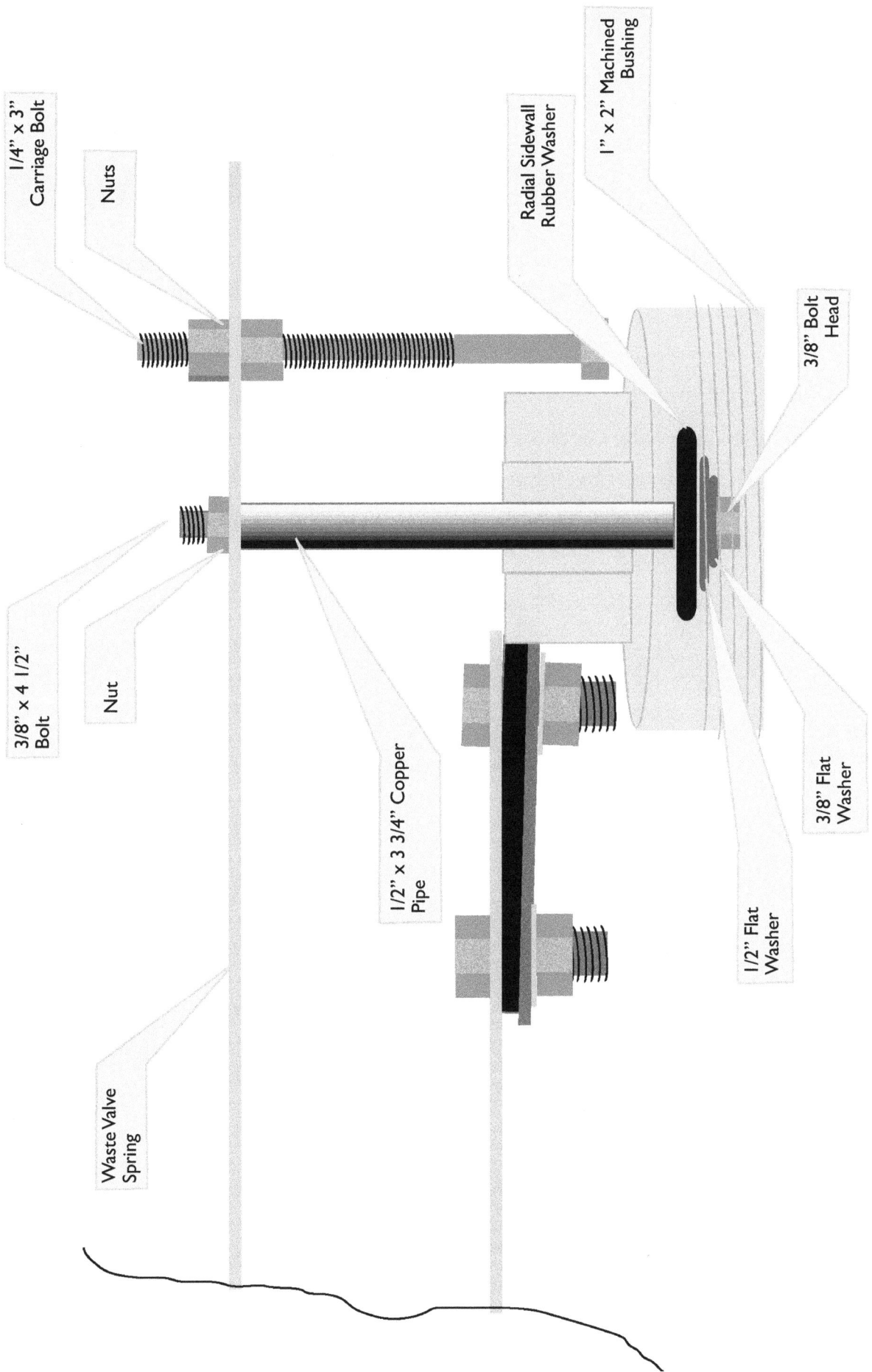

1/4" x 3" Carriage Bolt

Nuts

Radial Sidewall Rubber Washer

1" x 2" Machined Bushing

3/8" Bolt Head

3/8" x 4 1/2" Bolt

Nut

1/2" x 3 3/4" Copper Pipe

Waste Valve Spring

3/8" Flat Washer

1/2" Flat Washer

Figure 3-1: Plunger Assembly

Next, make the snifter hole or valve. Bore a 1/16" hole through the center of the 2" nipple that will go below the check valve. Once drilled, take some brass or copper wire and bend it to make a Cotter Pin (Figure 3-J). Place it through the hole, and then bend another right angle into the other end (Figure 3-J). Voila, it's pretty simple.

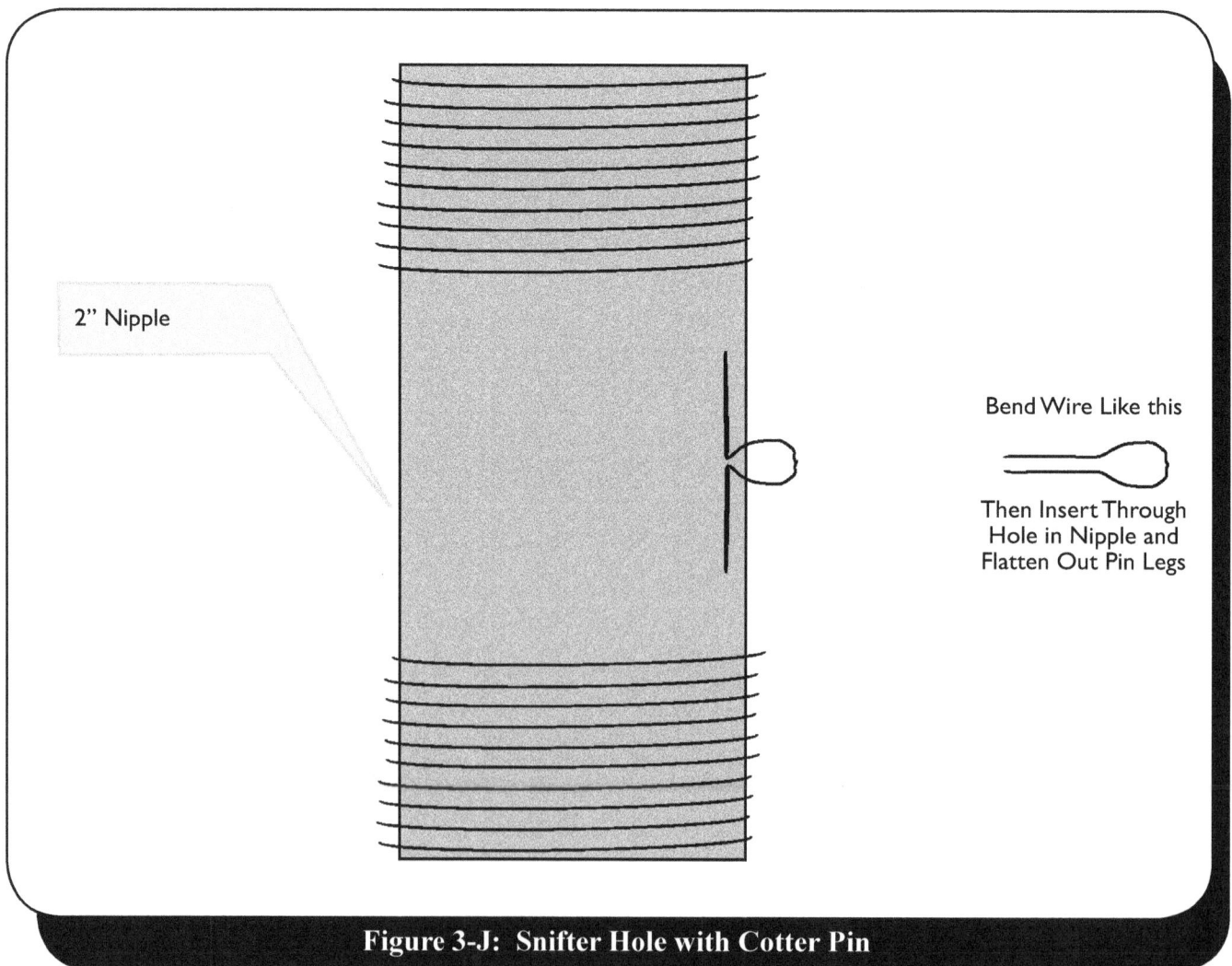

2" Nipple

Bend Wire Like this

Then Insert Through Hole in Nipple and Flatten Out Pin Legs

Figure 3-J: Snifter Hole with Cotter Pin

Putting It All Together

The rest of the construction is very simple. Follow diagram 3-A and screw each fitting into its proper place. Be sure to wrap all male threads with three wraps of plumber's Teflon tape. When wrapping, make a point to go in a clockwise direction if you're looking at the end of the pipe being taped so that the Teflon won't unravel when you begin to screw it into its female partner.

Also, all joints should be screwed together tightly, but not tremendously so. Snug them up well, but a pipe wrench large enough to grip 2" fittings will give you more leverage than you need for tightening. Be extra careful when wrapping and tightening the cap at the top of the air compression chamber. This fitting can be screwed on extra tight; it is holding air rather than water, so a leak will be difficult to detect.

A couple of tips for screwing in the waste valve assembly: make sure you screw it in place before putting the street bend into that tee so that the long spring can turn full revolutions. Also, DO NOT use the waste valve spring as a lever for tightening the waste valve into place. It is not a wrench and using it as one will bend the spring assembly and screw up all the delicate tweaking you've done to get the plunger centered in the bushing and the rubber clack closing perfectly. Use a pipe wrench and get a good, safe bite on the bushing to lever the waste valve around into a snug position on the tee.

Installation

Now that your pump is fully assembled it can be installed into your overall system as discussed in part II. Then 1" union just past the drive pipe gate valve makes attaching and removing the pump easy.

There is one other option I have added to our pump which is very helpful in tuning a ram when pumping significant vertical heights--a pressure gauge. By adding a 100 psi gauge on a 1/2" tee past the gate valve on the delivery line (see figure 3-A) you will be able to know how much pressure is on the system which is directly related to how much vertical height the water is being pumped. Many times I've worked on the pump here when the water table has dropped in the late Summer and the amount of available water to drive the ram to the height required is marginal. I used to adjust the ram, walk up the hill, check the delivery pipe, walk back down the hill, adjust the pump, etc. After installing a pressure gauge I can do all the tuning at the pump and know if the pump is ramming the water with enough pressure to reach the cistern at the top of the hill without climbing a hundred times. I will say, however, that the pressure gauges don't last long. They seem to quickly malfunction, probably a result of the constant bouncing back and forth of the needle on each stroke rather than a more gentle change of pressure as with the systems they're intended for.

When installing this pump there are only a few concerns to keep in mind. The air chamber should remain perpendicular to the ground. The waste valve needs to be below the check valve (a situation that is nearly impossible to goof up with this design). Finally, nothing should come into contact with the waste valve spring--it should be suspended in free air, not touching the wall of a pump house for instance.

This 2" pump is heavy enough for its size that it can simply rest on the ground, but it has a tendency to fall over if not somehow fixed in place. Here, I used a large galvanized electrical conduit clamps and small lag bolts to attach the ram to an oak slab

(Figure 3-K). A ram also can be anchored to a concrete slab if you provide anchor bolts in proper positions when pouring and if you fabricate metal straps to go over the pipes between bolts (Figure 3-L). Another very easy, extremely effective method of fixing the pump in place is to drive a metal fence post in the ground right at the base of the rear of the pump and attach the air chamber pipe to the fence post with pipe clamps (Figure 3-M). In some instances keeping the pump portable may be helpful, irrigating different areas of a farm at different times of the year, for example.

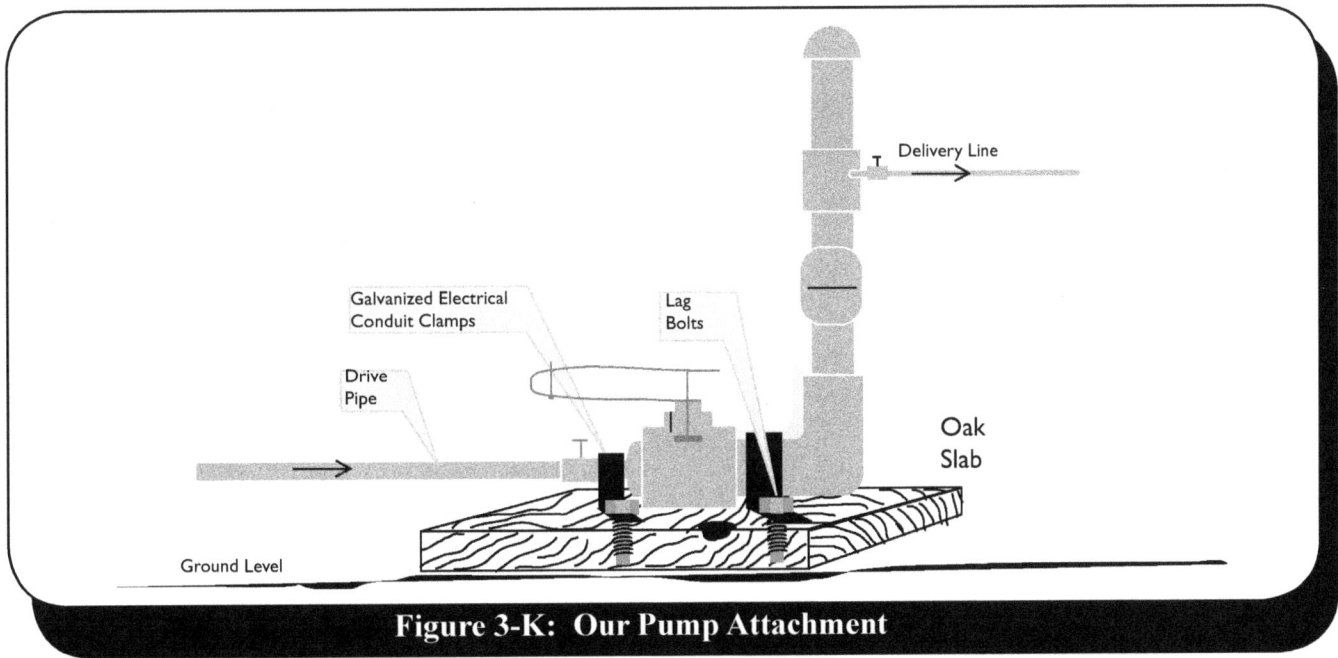

Figure 3-K: Our Pump Attachment

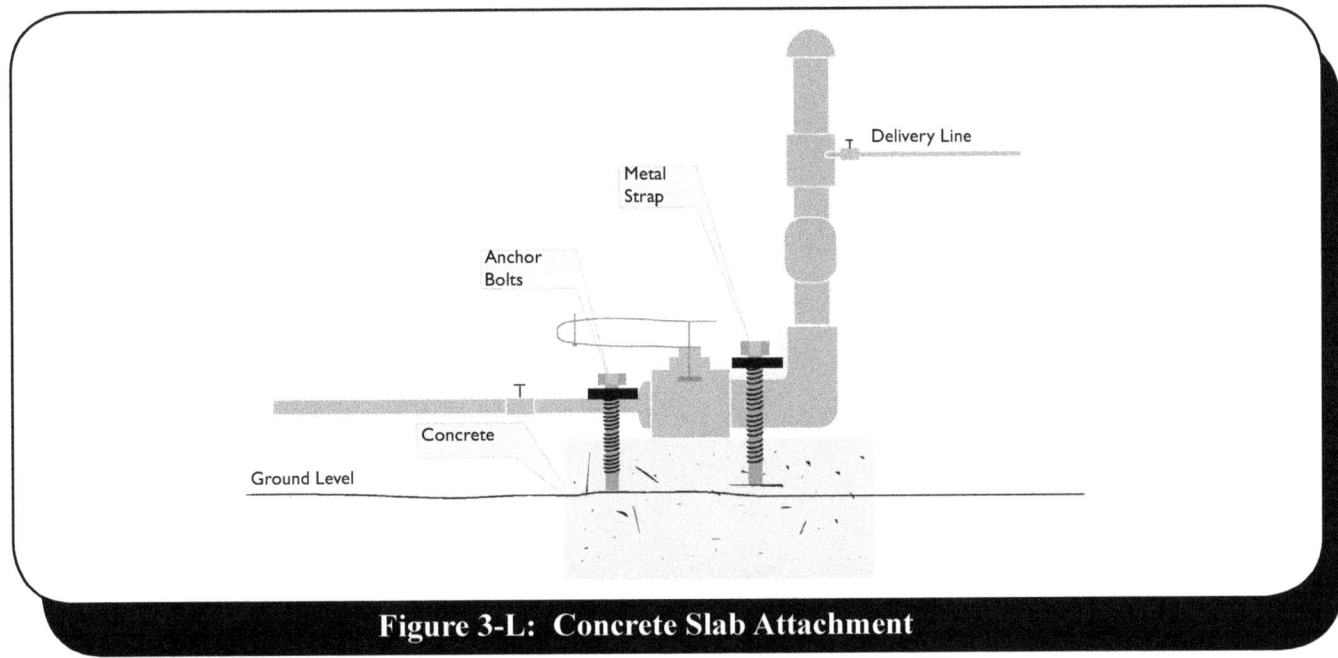

Figure 3-L: Concrete Slab Attachment

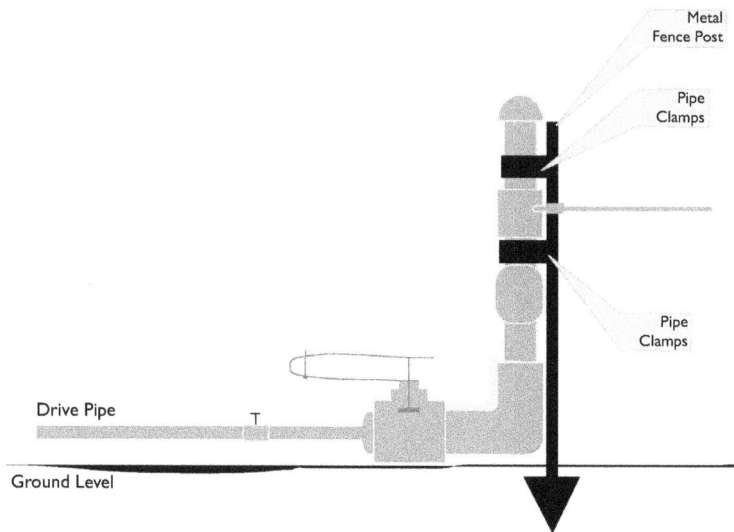

Figure 3-M: Fence Post Attachment

(Diagram labels: Metal Fence Post, Pipe Clamps, Pipe Clamps, Drive Pipe, T, Ground Level)

Getting It Running And Tuned Up

Many times when a ram is connected for the first time it will begin to run on its own. They seem to love to run. But, here's the whole rundown in order:

* Install drive pipe with gate valve at pump end, close gate valve so water isn't running

* Connect completed ram to end of drive pipe

* Connect delivery line gate valve and delivery line which should be run to water destination site and left unattached at the end so flow may be viewed once pump is running; delivery gate valve should be open

* Hold up on waste valve spring so plunger remains closed and then open drive pipe gate valve completely

* Water should enter the pump, flow through and up the delivery line to the height of the source, and a small spray should come from the snifter hole

* With the waste valve tension bolt completely loose and the carriage bolt stop completely up out of the way, let go of the waste valve spring which may or may not open on its own

* For proper operation, the drive pipe will need to be completely full of water, so you may need tinkering time to get air worked out of the drive pipe

* Slowly tighten the waste valve spring until the ram seems to operate smoothly; it will begin to open and close on its own. Note that as the water gets pumped upwards in the delivery pipe the pressure changes on the pump and will change its speed and overall operation until the water is being delivered all the way to the end site at which time pump rhythm should become constant

* Higher destinations require more pressure on the pump, so if it runs well but only delivers part way to the destination, increase waste valve spring tension

* If your available source water depletes and air is sucked into the drive pipe, the pump will stop running; adjust to a lighter spring tension and /or employ the carriage bolt stop by lowering it so it hits the top of the waste valve bushing preventing the waste valve from opening to its full capacity; NOTE: very tiny adjustments can make HUGE differences in pump function, so make very light tweaks

* Once running, the pump should cycle somewhere around 70 strokes per minute, but that's not the complete gospel. You'll get a feel for what your installation requires as you become familiar with it.

And that's it. This ram can provide years, even generations, of service. Usually the only problems encountered are a result of either a rubber piece in the pump wearing and needing replacement, or occasionally a cracked or broken bolt or metal fitting or a clogged snifter, but this is rare. The trouble-shooting section will help you solve various problems that can occur with this ram pump.

I hope you will find this technology helpful and a part of a larger quest to make environmentally sound decisions for providing for your needs.

PART FOUR

Troubleshooting

Ram Doesn't Run At All:

*   Manually open and close waste valve to begin stroking action
*   Spring tension not properly adjusted, adjust bolt on waste valve spring
*   Drive pipe not completely filled with water
*   Check for closed gate valves
*   Bad delivery check valve, unscrew air chamber to check
*   Rubber washer in waste valve not seating properly, disassemble waste valve and replace with newly cut washer
*   Hole somewhere in drive pipe
*   Blockage somewhere in drive pipe, evident if little or no water exits waste valve

Ram runs fine but doesn't deliver to destination

*   Not enough tension on waste valve spring to deliver to vertical height necessary, increase tension on waste valve spring bolt
*   Delivery check valve bad, remove air chamber to check
*   Air chamber water logged, close drive pipe and delivery pipe valves then remove delivery pipe (caution--air chamber is under pressure and water will spray
    when pipe is removed)
*   Leaky rubber washer on waste valve, disassemble and replace with newly cut washer
*   Leak in delivery line

Ram operated fine for some time, then quit

*   Ram consuming more water than available, getting air in drive pipe, adjust waste valve spring tension and/or waste valve stop bolt
*   Snifter clogged causing water logged air chamber
*   Blockage somewhere in drive pipe, evident if little or no water exits waste valve
*   Blockage in delivery pipe, disconnect delivery pipe from pump to see if water flows back down and out (caution--air chamber is under pressure and water will spray when pipe is removed)
*   Delivery or waste valves failed

Chattering

*   Chattering indicates air in the drive pipe

A P P E N D I X A

USA/UK/Metric
Conversions

Conversoin of Liquid Measurements:

To convert a quart to a liter multiply the quart measurement by .95. For the opposite, to go form liters to quarts, multiply the liters by 1.057. To convert gallons, to liters multiply by 3.7854.

USA measure:	UK	metric
1 fluid ounce	1.0408 UK fl oz	29.574 ml
1 pint (16 fl oz)	0.8327 UK pt	0.4731 l
1 gallon	0.8327 UK gal	3.78541 l

Conversion or Linear Measurements:

To convert inches into centimeters simply multiply the inch by 2.54. For the reverse, translating from centimeter to inches, multiply the centimeter measurement by .39. For instance, 2 inches are 5.08 cm.

imperial:	metric
1 inch [in]	2.54 cm
1 foot [ft] (12 in)	0.3048 m
1 yard [yd] (3ft)	0.9144 m
1 mile [mi] (1760 yd)	1.6093 km

Moates Homestead Water & Electrical Systems Schematic

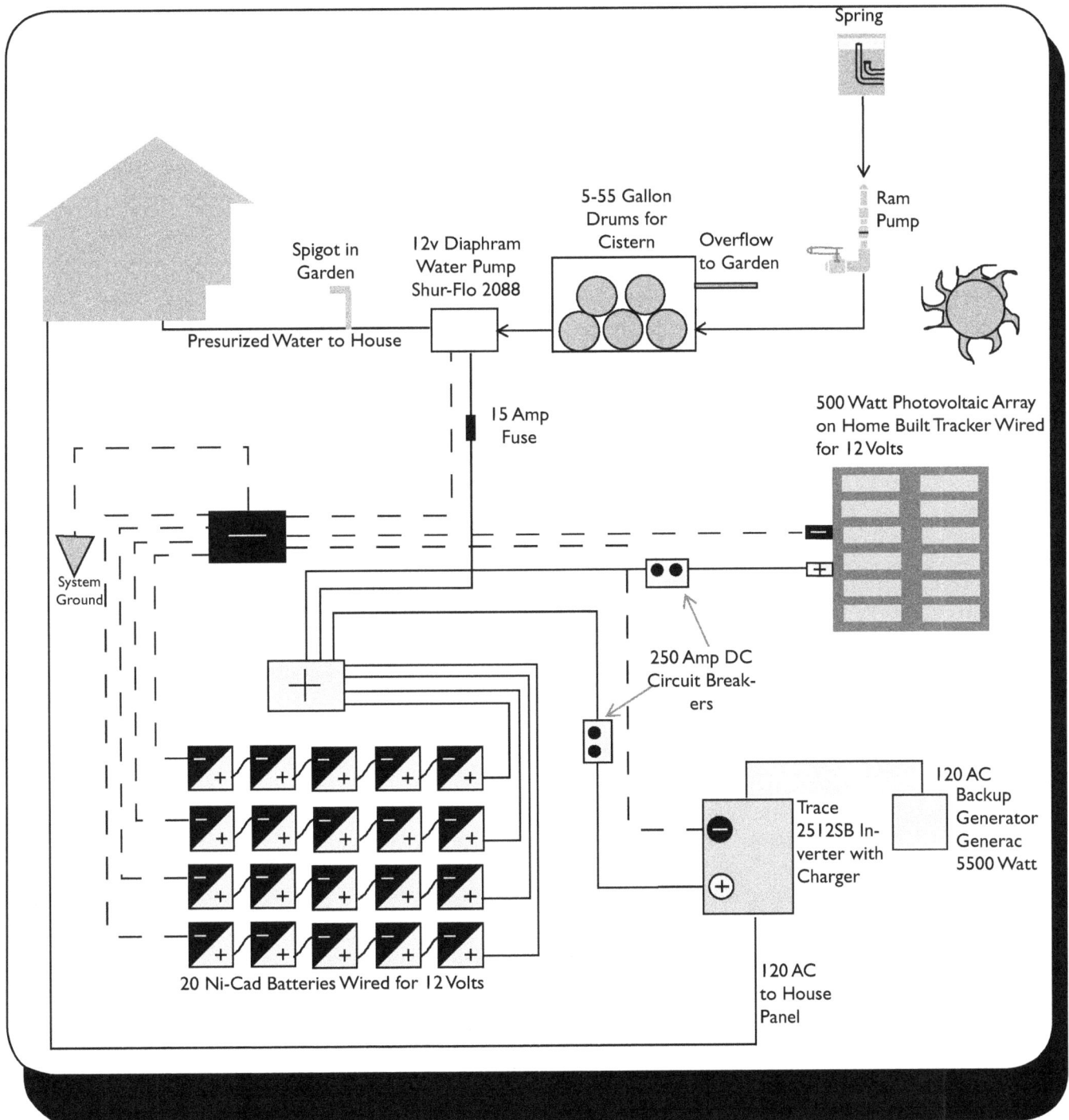

Spring

Ram Pump

5-55 Gallon Drums for Cistern

Overflow to Garden

Spigot in Garden

12v Diaphram Water Pump Shur-Flo 2088

Presurized Water to House

15 Amp Fuse

System Ground

500 Watt Photovoltaic Array on Home Built Tracker Wired for 12 Volts

250 Amp DC Circuit Breakers

20 Ni-Cad Batteries Wired for 12 Volts

Trace 2512SB Inverter with Charger

120 AC Backup Generator Generac 5500 Watt

120 AC to House Panel

www.ingramcontent.com/pod-product-compliance
Lightning Source LLC
Chambersburg PA
CBHW061417090426
42742CB00027B/3501